Theory and Application of Acoustic Sources Using Complex Analysis

Martin Ochmann · Rafael Piscoya

Theory and Application of Acoustic Sources Using Complex Analysis

Complex Acoustic Sources, Green's Functions and Half-Space Problems, Acoustic Radiation and Scattering Using Equivalent Source and Boundary Element Methods

Springer

Martin Ochmann
Project Group Computational Acoustics
Emeritus of Beuth University of Applied Sciences
Berlin, Germany

Rafael Piscoya
Project Group Computational Acoustics
Beuth University of Applied Sciences
Berlin, Germany

ISBN 978-981-33-6039-6 ISBN 978-981-33-6040-2 (eBook)
https://doi.org/10.1007/978-981-33-6040-2

This Springer imprint is published by the registered company Springer Nature Singapore Pte Ltd.
The registered company address is: 152 Beach Road, #21-01/04 Gateway East, Singapore 189721, Singapore

Preface

The sources of sound are the most important elements of acoustics. Without sources, there will be no sound at all. Sources generate all sound fields, the desired ones like music and the unwanted ones like unpleasant and disturbing noise.

Point sources like monopoles, dipoles, etc. are the simplest acoustical sources that play a prominent role in technical and numerical acoustics, since they can be used to construct more complex sources by linear superposition. Also, they are keystones in numerical methods like boundary element methods and equivalent source methods.

Such point sources possess a singularity at the source location from which the sound—figuratively speaking—flows out. This singularity is usually located in the three-dimensional real physical space. However, when dealing with the topic of sound propagation over impedance planes, it turns out that the sound field can only be assembled by point sources with singularities located in the complex plane, in order to ensure convergent mathematical expressions for all kind of surface impedances. That was really a surprise and raised a number of questions: Is this just a mathematical trick of complex analysis to shift the line of integration into the complex domain? Or does this "complex" monopoles or multipoles have its own physical meaning? The mathematical source of a real monopole is a Dirac delta function on the right-hand side of the Helmholtz equation. Does it make sense to define a Dirac delta function with complex argument, in order to generate a complex monopole? These questions should be studied and answered in the present book. In addition, numerous applications of complex sources will be shown. It will be demonstrated how such sources can be practically realized.

By combining acoustical source theory with complex analysis, a large and rich field of research opens up with many still unsolved problems and unanswered questions and with fascinating mathematical and physical structures.

The aim of this book is to present some of our research results from the last 20 years (approx. 2000–2020) in a comprehensive and hopefully clear form. The reader will decide whether this attempt has been successful or not.

Large parts of the book are published here for the first time, and therefore, it offers suggestions for further research. For this reason, the book is intended primarily for researchers in the field of theoretical and numerical acoustics, but for technical acoustics, too, since applications of complex sources are also considered.

Some chapters contain many formulas, but we hope that the reader can see the most simple idea behind the equations. It was our intention to give detailed derivations of the solutions obtained for the convenience of the reader.

Clearly, we are aware that nobody is immune to mistakes and thus, we would be grateful to readers, who draw our intention to errors, whenever they have found them.

Although the book is a common work of both authors based on a long and fruitful cooperation in many research projects during the nearly two decades, we would like to mention that Chaps. 1–6 come from the pen of Martin Ochmann and Chaps. 7–11 were written by Rafael Piscoya.

Berlin, Germany

Martin Ochmann
Rafael Piscoya

Acknowledgements

The research project *"Theory and application of acoustical multipole radiators with complex singularities"* (OC 16/9-1 and OC 16/9-2) started in 2015 and was funded by the German Research Foundation (DFG) for five years.

Thus, the authors gratefully acknowledge the financial support of the DFG within this research project without which the book could not have been written.

We also would like to thank the Senior Editor of Springer Nature in Singapore Dr. Loyola D'Silva and his team for his support. I (Martin Ochmann) met him at the Euronoise conference 2018 in Crete, Greece, where Loyola asks me to publish a book about our newest research results. That was the birth of the present book.

Contents

Symbols (for Chaps. 2–6)

c	Sound speed
$D_1,\ D_2,\ D_3$	Dipole contribution in 1D, 2D, and 3D
$F_x(f)$	Fourier transform in x of f
$F_x^{-1}(F)$	Inverse Fourier transform of F
$g_1,\ g_2,\ g_3$	Complex monopole in 1D, 2D, and 3D
$G_2,\ G_3$	Half-space Green's function in 2D and in 3D
H	Heaviside unit step function
$H_0^{(1)}$	Hankel function of the first kind and order zero
$H_r(f)$	Hankel transform of order zero of f
$H_r^{-1}(F)$	Inverse Hankel transform of order zero of F
i	Imaginary unit
Im{ }	Imaginary part
$j_0(z)$	Spherical Bessel function of order zero
$J_0(z),\ J_1(z),\ J_{1/2}(z)$	Bessel functions
k	Wave number (= ω/c)
$M_1,\ M_2,\ M_3$	Monopole contribution in 1D, 2D, and 3D
n	Unit surface normal
p	Sound pressure
r	$\sqrt{x^2+y^2}$
$\tilde{r}$	Complex distance
$\vec{r}_c$	Vector of complex distance $(= (x, y, z+ia)^T)$
R	Reflection coefficient
$\mathbb{R}^3$	Field of real 3D numbers
Re{ }	Real part
$S_1,\ S_2,\ S_3$	Driving sources in 1D, 2D, and 3D
sgn(z)	Signum function
$u_0(z)$	"Basis" solution in free space
$V_+,\ V_-$	Upper and lower half-space
v_n	Normal velocity

z_h	$z + h$
Z	Impedance
Z_0	Normalized acoustical impedance ($= Z/(\rho c)$)

Greek

γ	ik/Z_0
δ	Dirac delta function
$\tilde{\delta}_1,\ \tilde{\delta}_2,\ \tilde{\delta}_3$	Driving sources in 1D, 2D, and 3D
$\Delta,\ \nabla$	Laplacian and Nabla operator
(ζ, ξ, ϕ)	Oblate spheroidal coordinates
μ	$= \sqrt{1 - 1/Z_0^2}$
ρ	Fluid density
$\sigma_1^{(1)},\ \sigma_2^{(1)},\ \sigma_3^{(1)}$	Correction terms for the sound rigid case with real sources
$\sigma_1^{(2)},\ \sigma_2^{(2)},\ \sigma_3^{(2)}$	Correction terms for the sound rigid case with complex sources
$\phi(\vec{r}_c)$	Simplified complex monopole
$\chi_{(-a,a)}(x)$	Characteristic function of the interval (–a,a)
ω	Angular frequency

Chapter 1
Introduction

1.1 Motivation

During a workshop on Physical Acoustics of the German Acoustical Society in Bad Honnef, Germany in the year 1999, Prof. F. P. Mechel gave a lecture on the fundamentals of mirror and corner sources (title in German: *Grundlagen des Spiegelquellen- und des Kantenquellen-Modells*; Abstract in English: [Mec00a]). In his talk, Mechel mentioned that "*the mirror source is an exact solution of the reflection at infinite rigid walls, but it is only approximation for absorbing walls.*" The reason is that the usual simplified construction refers to the scattering of plane waves. It consists of the mirrored point source multiplied by a spatially dependent reflection factor which together does not satisfy the wave equation. This aroused the interest of one of the authors (Martin Ochmann) to look for exact solutions. First, he came across the original "ancient" approach of Arnold Sommerfeld (see [Som78], Chap. 13), where a superposition of point sources in form of a line integral below the plane for treating problems of heat conduction is used. This solution could be transferred to the related acoustical problem of a point source above an impedance plane without problems. But, surprisingly, it turned out that the line integral of mirror sources only converges for mass-like or pure absorbing surface impedances at the reflecting plane. For all other kinds of impedances, the line integral diverges and no meaningful solution could be found. What is the reason for this strange behavior? After many unsuccessful attempts, it was found that the line integral converges for an arbitrary surface impedance, if the point sources are allowed to have complex locations (see [Och04]). Mathematically speaking, the path of integration must be shifted from the real to the imaginary axis. Similar approaches could now be found in the literature, especially in the electromagnetic field.

However, another question arose: Are the sources with complex position a purely mathematical construction, namely as an application of the theory of the analytical continuation of functions into the complex plane, or do they have their own physical meaning?

M. Ochmann and R. Piscoya, *Theory and Application of Acoustic Sources Using Complex Analysis*, https://doi.org/10.1007/978-981-33-6040-2_1

Scanning through the relevant literature, we learned that such complex sources—like almost everything—had been investigated earlier. However, in acoustics such source constructions were nearly unknown and were studied more often in electromagnetic wave theory. Thus, complex sources turned out to be strange physical objects of independent importance and should therefore be examined in their own right—detached from the original problem of sound propagation about impedance planes.

1.2 Rough Overview and Structure of the Book

In Chap. 2, we start with the definition and discussion of the simple source, the point source, or the monopole in three dimensions (3D) and add an imaginary part to the location or singularity of the source. Such a source with a complex location will be called a "complex source" or a "complex monopole." We study some basic properties of the complex monopole like the fact that the singularity of the source now becomes extended in space and forms a singularity disk. It is shown that the complex monopole is—just as the real monopole—a solution of an inhomogeneous Helmholtz equation. However, at this point, the inhomogeneity of the Helmholtz equation, the so-called driving source, is still unknown.

In the short Chap. 3, it is shown that the complex monopole in oblate spheroidal coordinates takes a very simple shape. However, this representation is no longer used in the following, since we will mainly work in Cartesian coordinates.

Now we know that the complex monopole satisfies the reduced wave equation or Helmholtz equation outside the singularity disk. This immediately leads to the central question of which inhomogeneity is responsible for the source singularity. In other words: What is the mathematical source of the complex monopole? The answer is more complex than one would initially assume. It is not a Dirac delta function with complex coordinates. The solution will be given in Chap. 4 by starting with the one-dimensional case (1D). It will turn out that the 1D driving source can be found relatively easily and that it will be the decisive key for constructing the driving sources in 2D and in 3D by using the Sommerfeld or Weyl integrals.

We are now equipped with enough knowledge about complex monopoles to turn to an important application of complex sources. In Chap. 5, the half-space Green's function above impedance planes in 1D, 2D, and 3D is determined by superposing real or complex multipoles in a suitable way, which brings us back to the motivating idea of the book as described above. For that reason, a general impedance formula for the superposition integral of real and complex sources (Theorem 4) is derived and presented for the first time. Based on this formula, it is now possible to obtain expressions for the half-line, half-plane, and half-space Green's function together with its corresponding driving sources. Also, a generalized formula for all three driving sources depending on the dimension is found (see Eq. 5.89).

Until now we have used the solution of the half-space problem for finding the driving source terms. Knowing this, we can do the opposite and try to solve the corresponding inhomogeneous Helmholtz equation in free space instead of considering the original boundary value problem. Such an approach will be used in Chap. 6 to find old and new solutions of the half-space problem in 2D and in 3D by means of the Helmholtz Huygens Integral and by a combination of Fourier and Hankel transform techniques. By this procedure, we will obtain again the solutions we already know, which confirms the correctness of the theory and the calculations. The newly derived formulas complement the available solutions to the half-space problem.

While the first part of the book focuses mainly on the theory, the second part deals more with practical implementations of this theory. The half-space Green's function has several integral representations that have terms containing square roots of complex functions. In order to evaluate correctly the integrals, the proper branch cuts need to be selected to ensure having continuous integrands. In Chap. 7, different branch cuts for the complex distance function and the square root for the Sommerfeld integral are studied.

In the next Chap. 8, the feasibility of a physical realization of complex sources is investigated, since the question of the physical existence of such sources is inevitable. It is shown that loudspeaker arrays with specific characteristics can reproduce the radiation patterns of complex multipoles.

When the solution of practical problems need to be obtained with proper accuracy, theoretical approximations are usually not enough and one has to use computational methods such as the equivalent source method (ESM), the boundary element method (BEM), or the finite element method (FEM). Chap. 9 presents an extension of the ESM by including complex monopoles. The idea is to take advantage of the focusing of the complex monopoles to reduce the total number of equivalent sources. Several numerical examples are given to evaluate the performance of the modified method.

Up to this point, the absorbing ground has been characterized by a locally reacting impedance. In Chap. 10, a homogeneous ground with frequency independent density and sound speed is considered. In 2D, only the reflected wave is analyzed. It is shown that it can be represented by integrals with the integration path lying in the complex plane. In 3D, the problem can be solved by finding complex equivalent source distributions following the same approach as in electromagnetics. This approach, called exact image theory (EIT), is verified by a comparison with the solution using the Sommerfeld integrals.

In the last Chap. 11, another well-known computational approach, namely the BEM, is described and its half-space formulation is explained. Four different representations of the Green's function, including the one in terms of complex monopoles, are analyzed; their numerical stability is evaluated; and their performance by a monopole test is compared. An example of application is given, where the insertion loss of a sound barrier is calculated.

Finally, we would like to remind the reader, that the denotation for the imaginary unit $i = \sqrt{-1}$ together with the time convention $\mathrm{e}^{-i\omega t}$ is used in Chaps. 2–6 while the denotation $j = \sqrt{-1}$ together with the time convention $\mathrm{e}^{+i\omega t}$ is used in Chaps. 7–11 (except in the section about the EIT).

1.3 Historical and Scientific Development of Complex Sources and Their Application

In the first volume of his work about sound absorbers ([Mec89], in German), Mechel wrote a 154-page overview article entitled "*Spherical waves over an absorbing plane,*" in which he derives and discusses many of the previously known solutions of spherical waves over homogeneous and locally reacting absorbers.

As Mechel mentioned (see [Mec89], p. 522), "*everything started with*" Arnold Sommerfeld ([Som09, Som26, Som78]) who investigated in the year 1909 the electromagnetic field of a point source in a half-space bounded by an infinite plane with extended reaction, also called homogeneous ground.

Since that time the problem of determining the corresponding half-space Green's function of the associated acoustical problem was treated by many researchers. Such a half-space Green's function can be incorporated in a BEM (see [Och05a, Och08a, Och08c, Bri09, Och13c, Pis17, and Pis19]) making the BEM much more effective, since the infinite plane itself has not to be discretized.

As detailed in the *motivation*, Sect. 1.1, half-space Green's functions can be composed of complex point sources. Such "*complex source point solutions*" [Fel76a] have been studied in the electromagnetic literature for a long time. First papers are given by Izmest'ev [Izm70] and Deschamps [Des71] and motivated numerous investigations in which such exact solutions of the wave equations and their connection with decaying, evanescent waves and with Gaussian beams were considered (see, e.g., [Fel76b, Zio85, Zau86, Hey89, Kai03]). A review article with over 100 references to localized light waves and the associated exact solutions were published by Kiselev in 2007 [Kis07]. Therein, complex monopoles are called "localized exact solutions" or "spherical waves complexified by means of shift," since Kiselev noted that the term "*field of a complex source ... was but a metaphor*" (p. 616 of [Kis07]). As already mentioned above, the first applications of the "Complex-source-point solutions" in electromagnetic research can be found to solve the famous, classical Sommerfeld problem ([Som09, Som26, Som78], see also [Moh82]), in which Hertzian dipoles are studied over conductive ground. The so-called complex image theory was then developed in order to calculate the field of antennas in half-spaces above conductive ground. Some works are listed here as examples: [Fel94, Lin84, Lin85, Lin86, Yan91, Shu93, Vit96, and Pir98]. Electromagnetic standing waves in resonators can also be represented by superimposing complex sources [Cul79].

The application of complex sources in the electromagnetic field theory had an impact on theoretical acoustics. Complex monopole radiators were used under the name "*method of complex images*" to calculate the sound propagation above impedance ground ([Di93, Li96, Och04, Tar05a, Tar05b, Och08, O'Ne12, Och13a,b]) and above seabeds ([Faw00, Faw03a, Faw03b]). Norris and Hansen showed that the sound field radiated from compact source regions could be presented by distributions of complex point sources [Nor97]. In this context, they were able to show that continuously distributed monopoles that are located on a "complex sphere" can reproduce real multipoles. The relationship with the theory of Gaussian beams has already been considered in the earlier work [Nor86]. In [Cou81], it is even proven that the paraxial Gaussian beam passes over into the complex monopole when all terms of an associated perturbation calculation are taken into account.

If the monopole source moves with constant speed parallel to the impedance plane, it was shown that the corresponding solution for the resulting sound field consists of monopoles and dipoles with complex source locations [Och13a]. Therefore, the question naturally arises whether it is possible and useful to consider general multipoles with complex singularities and to extend the "*source point complexification*" of the zeroth- (monopole) and first-order (three dipoles) point radiators on general point sources of higher order. For the sake of simplicity, such sound fields will be referred to below as complex monopoles, dipoles, etc. Complex multipoles are rarely studied in the literature. In electromagnetics, three papers were found in which complex multipoles are treated or briefly mentioned: In the short paper by Shin and Felsen [Shi77], complex multipoles are defined by spatial differentiation of the complex monopole. Izmest'ev proposed in [Izm70] to use oblate spheroidal coordinates (see Chap. 3). But he gives only asymptotic formulas for a possible representation of complex multipoles. Zauderer [Zau86] refers on the definition of complex multipoles given in [Shi77] and shows that the Hermite-Gaussian and Laguerre-Gaussian beams with a complex argument are paraxial approximations of complex multipoles. Zauderer therefore also regards these functions as derivatives of complex monopoles, following Maxwell's theory of multipoles (see, e.g., [Som78], p. 140 ff.). An explicit connection with the usual multipole representations in spherical coordinates is unfortunately not given. In the present book, our focus is on complex monopoles, but Sect. 8.5 gives a short introduction into the theory of complex multipoles.

A helpful tool for the analysis and synthesis of sound fields is the integral representations by Weyl and Sommerfeld, which develop real point sources in plane or cylindrical waves (see, e.g., [Duf01]). Li and White (see [Li96], p. 757) and Kaiser ([Kai03], R319 ff.)) investigated whether these integral representations could be extended to complex monopoles. While Li and White restrict themselves to a "simple numerical test," Kaiser succeeds in proving a "*generalized Weyl formula*" for the complex monopole in 3D, which he called time-harmonic complex beam. In Sect. 4.2, we give a proof for the analogous "*generalized Weyl or Sommerfeld formula*" in 2D (see Eq. 4.17).

The sound fields generated by complex sources are characterized by distinct directivity patterns (see Chap. 8), which in turn are determined by the driving sources or, in other words, by the right-hand sides of the Helmholtz equation. The reader should note that the term "*source*" has different meanings depending on the context. This is explained in more detail in Sect. 4.1. Kaiser examines in a series of papers [Kai00, Kai01, Kai03] these associated driving sources, which he calls "*extended delta functions*" or "*singular source distributions.*" Further related literature can be found in the review article [Kai03]. Kaiser derives an explicit representation for the n-dimensional potential equation and discusses his result for the special case of three-dimensional space (see [Kai00], p. 14, Proposition 3). This is a very promising approach to derive analogues for the three-dimensional Helmholtz equation. Such source representations are derived by Tagirdzhanov et al. in [Tag09, Tag011, Tag013, and Tag015] by using another approach (see [Tag11], p. 2). Surprisingly, the structure of the driving sources depends on the choice of the branch cut. In his work, Kaiser explains important aspects of such source distributions, which are no longer limited to a point-wise singularity, but rather "*become 'inflated' to extended objects*," and he states: "*The real coordinates give the center, and the imaginary coordinates the extent and orientation of the object in space and time*" ([Kai03], quotation from p. R293). A wealth of tools from mathematical physics for the analysis of complex point sources is discussed in this work, such as, e.g., branch cuts, wavelets, disk sources, oblate spheroidal coordinates, hyperfunctions, holomorphic potentials, and Riemann functions. In [Hey00], Kaiser's theory of complex sources is used to describe acoustical sources and fields in the frequency and in the time domain.

As mentioned above, complex multipoles can be used to numerically construct sound fields with a given directional characteristic (see Chap. 8). Such sound beams, also called audio spotlight [Yon83], have been generated in different ways. A method that has long been known comes from nonlinear acoustics and is called "*parametric array*" [Wes63]. This approach makes use of the fact that high-energy acoustic waves of neighboring frequencies interact with one another and, due to the nonlinearity, generate strongly directed difference tones. Mellert and Schwarz-Röhr [Mel04] used this nonlinear interaction for producing experimentally low-frequency directed sound beams. There are already various industrial products and applications on the market. The dissertation [Ols08, in German] gives an overview of the theory and realization of such strongly directed sound beams in the audio range using parametric ultrasonic loudspeakers. However, the equations of nonlinear acoustics are known to be very difficult to work with. Thus, complex multipoles could offer a theoretically simpler alternative for modeling strongly focused sound fields, since only linear theory is needed.

A well-known method for calculating the sound radiation from vibrating structures is the ESM, also known as the source simulation technique. The ESM mostly uses multipole sources with real singularities as elements of the equivalent source system, which is supposed to replace the sound-radiating body by adapting the given vibrating velocity on the boundary of the radiating body. An overview of the ESM with further references can be found in [Och95, Och00, Och08a, Och08b]. It can be shown that the ESM equations have the disadvantage compared to boundary integral equations that they lead to ill-conditioned systems. The corresponding linear systems of equations resulting from discretization are characterized in almost all cases by very high condition numbers, which can lead to unstable numerical calculations. These relationships were studied in [Och99]. Alternative equivalent sources such as area-like sources that are weighted with a Gaussian function [Guy94], or descriptions in prolate spheroidal coordinates [Hac84] were also used. These approaches already indicate the use of *equivalent complex sources*, since they possess spatially extended singularities and have a simple shape in spheroidal coordinates (see Chap. 3) The idea of integrating complex monopole sources into the source system in order to increase the efficiency and stability of the method was carried out for the first time in [Och06], in order to calculate the sound field radiated from a circular baffled piston. The extension of the singularity from a point (real source) to a disk (complex source) represents a limitation in some cases (since the sources must be located completely in the radiating body) but the presence of a focused radiation can bring a reduction in the total number of sources. More details and different cases of application can be found in Chap. 9 of the present book.

Another example of application is the modeling of jet engines by ring sources on the basis of measurement data [Hol94]. Since the singularities of a complex monopole lie on circles, equivalent complex sources seems to be well suited for modelling rotationally symmetrical radiators.

If the half-space Green's function is known—either above homogeneous ground (see Chap. 10) or above impedance ground (Chap. 11)—it can be incorporated in a BEM (see [Och08c, Pis17, Pis19]), and consequently, there is no need to model the infinite plane itself, too. For this purpose, the expression for the half-space Green's function should be as simple as possible, in order to lead to a fast numerical method, especially if the radiation or scattering from complicated structures consisting of many thousand boundary elements is considered. From the several formulas for the half-space Green's function that can be found in this book, the formula containing complex monopoles is the easiest to implement, since the integrand has simply the term e^{ikr}/r with the only difference that r is complex. This representation serves as a reference for the comparison with other representations as can be found in Chap. 11.

In the following Chaps. 2–6, *the imaginary unit is denoted by* $i = \sqrt{-1}$ *and the time convention* $e^{-i\omega t}$ *is used.*

Chapter 2
Complex Monopoles and the Helmholtz Equation in Cartesian Coordinates

Consider the slightly simplified complex monopole in Cartesian coordinates (*x*, *y*, *z*)

$$\varphi(\vec{r}_c) = \varphi(x, y, z + ia) = \frac{e^{ik\sqrt{x^2+y^2+(z+ia)^2}}}{\sqrt{x^2 + y^2 + (z + ia)^2}} = \frac{e^{ik\tilde{r}}}{\tilde{r}} \tag{2.1}$$

for $\vec{r}_c = (x, y, z + ia)^T$, $\tilde{r} = \sqrt{x^2 + y^2 + (z + ia)^2}$ with a real constant *a*. Here, the superscript *T* denotes transposition, (x, y, z) is the receiver position, $k = \omega/c$ is the wave number with angular frequency ω, and *c* is the speed of sound. All time-varying quantities should obey the time dependence $e^{-i\omega t}$ with $i = \sqrt{-1}$.

All properties, which will be derived in the sequel, are also valid for the more general case, where the complex distance is given by

$$\tilde{r} = \sqrt{(x - ia)^2 + (y - ib)^2 + (z - ic)^2} \quad (a, b, \text{and}\, c \text{ are real constants}). \tag{2.2}$$

A discussion of the complex monopole in its general form can be found at our website—https://projekt.beuth-hochschule.de/?id=2825—and Chap. 7 where possible branch cuts of the square root are explained.

We will show that φ is a solution of the Helmholtz equation with respect to the spatial coordinates (*x*, *y*, *z*) for the fixed complex source point –*ia*. Here, the complex distance becomes zero on the circle

$$C = \left\{(x, y, z)\middle|x^2 + y^2 = a^2, z = 0\right\} \tag{2.3}$$

around the *z*-axis at $z = 0$ with radius *a*. Furthermore, the complex root must be made single-valued. It can be shown (see [Kai00]), that a suitable branch cut, requiring that $\mathrm{Re}\tilde{r} \geq 0$, is given by the disk

$$D = \left\{(x, y, z)\middle|x^2 + y^2 \leq a^2, z = 0\right\}. \tag{2.4}$$

M. Ochmann and R. Piscoya, *Theory and Application of Acoustic Sources Using Complex Analysis*, https://doi.org/10.1007/978-981-33-6040-2_2

Hence, the complex monopole (2.1) becomes singular on the source disk *D*. Outside of *D*, the complex source is an exact solution of the Helmholtz equation. This can be shown as usual, since the proof is the same whether the source point is complex or real:

Proof Let $\nabla = (\partial/\partial x, \partial/\partial y, \partial/\partial z)^T$ be the gradient and $\Delta = \nabla \cdot \nabla$ (the point denotes the scalar product) the Laplacian with respect to (*x*, *y*, *z*) for *a* fixed. Then, we obtain

$$\nabla\varphi = \nabla\left(\frac{e^{ik\tilde{r}}}{\tilde{r}}\right) = \mathrm{e}^{ik\tilde{r}}\nabla\frac{1}{\tilde{r}} + \frac{1}{\tilde{r}}\nabla\mathrm{e}^{ik\tilde{r}} \Rightarrow \Delta\varphi = \mathrm{e}^{ik\tilde{r}}\Delta\frac{1}{\tilde{r}} + 2\nabla\frac{1}{\tilde{r}}\cdot\nabla\mathrm{e}^{ik\tilde{r}} + \frac{1}{\tilde{r}}\Delta\mathrm{e}^{ik\tilde{r}}. \tag{2.5}$$

The first term is zero, since the potential $\vartheta = 1/\tilde{r}$ is harmonic with respect to (*x*, *y*, *z*) outside the disk *D* (see Proposition 1 of [Kai00]).

The second term gives

$$2\nabla\frac{1}{\tilde{r}}\cdot\nabla\mathrm{e}^{ik\tilde{r}} = -2ik\mathrm{e}^{ik\tilde{r}}\frac{1}{\tilde{r}^2}, \tag{2.6}$$

since $\nabla\frac{1}{\tilde{r}} = -\frac{1}{\tilde{r}^2}\nabla\tilde{r}$, $\nabla\mathrm{e}^{ik\tilde{r}} = ik\mathrm{e}^{ik\tilde{r}}\nabla\tilde{r}$, and $\nabla\tilde{r}\cdot\nabla\tilde{r} = 1$, where we have used that

$$\nabla\tilde{r} = \nabla\sqrt{x^2+y^2+(z+ia)^2} = \frac{\vec{r}_q}{\tilde{r}}.$$

The last term on the right-hand side of Eq. (2.5) can be written as

$$\frac{1}{\tilde{r}}\Delta\mathrm{e}^{ik\tilde{r}} = \frac{\mathrm{e}^{ik\tilde{r}}}{\tilde{r}}(-k^2 + 2ik/\tilde{r}) = \varphi\left(-k^2 + \frac{2ik}{\tilde{r}}\right), \tag{2.7}$$

since $\Delta\mathrm{e}^{ik\tilde{r}} = \nabla\cdot\nabla\mathrm{e}^{ik\tilde{r}} = \nabla\cdot\left(ik\mathrm{e}^{ik\tilde{r}}\nabla\tilde{r}\right) = \mathrm{e}^{ik\tilde{r}}\left(-k^2 + ik\Delta\tilde{r}\right)$ and $\Delta\tilde{r} = \frac{2}{\tilde{r}}$.

By combining Eqs. (2.5), (2.6), and (2.7), we immediately get

$$\Delta\varphi + k^2\varphi = 0 \quad \text{for } (x, y, z) \notin D. \ \square \tag{2.8}$$

The symbol □ denotes the end of the proof. But what does such a complex monopole look like? For positive values of the imaginary *z*-location it behaves in a similar manner as surface waves. Whereas the ordinary monopole source becomes singular at a single point (point source), the complex point source shows a different behavior: It becomes singular on disks enclosed by circles around the *z*-axis at $z = 0$ with radius *a* (circle or ring sources), and for $|a| \to \infty$ the radius of these circles increases more and more like a growing circular surface wave. Figure 2.1 shows the development of a complex point source with increasing positive values of *a*.

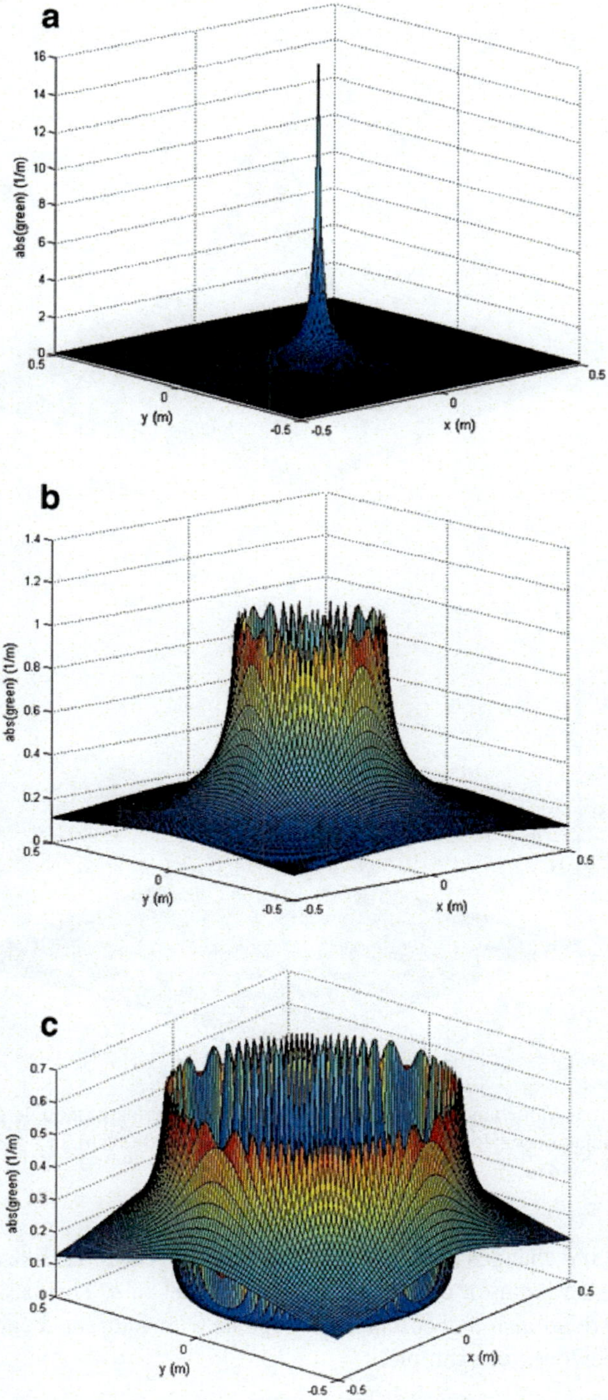

Fig. 2.1 Absolute value abs(green) $= |4\pi\varphi(\vec{r}_c)|$ of the complex monopole (2.1) for increasing values of the imaginary z-location a over the xy-plane; $z = 0.005\,\mathrm{m}$ and $f = 1000\,\mathrm{Hz}$; **a** $a = 0\,\mathrm{m}$, **b** $a = 0.2\,\mathrm{m}$, **c** $a = 0.4\,\mathrm{m}$

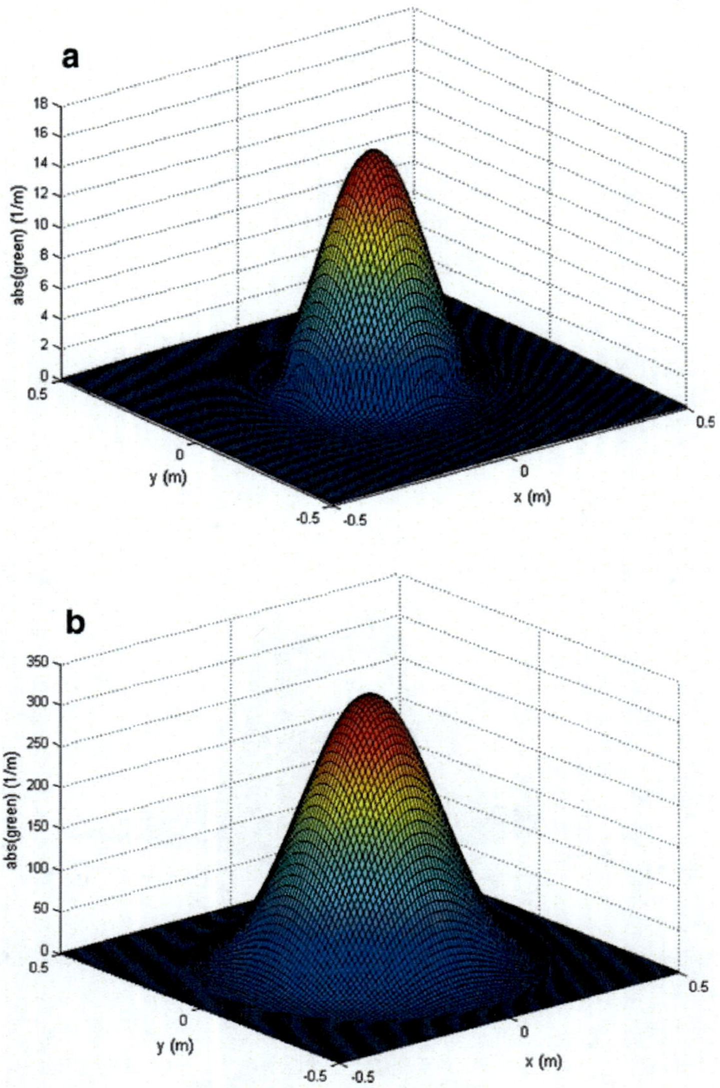

Fig. 2.2 Absolute value abs(green) $= |4\pi\varphi(\vec{r}_c)|$ of the complex monopole (2.1) for decreasing values of the imaginary z-location a over the xy-plane; $z = 0.005$ m and $f = 1000$ Hz; **a** $a = -0.2$ m, **b** $a = -0.4$ m

However, if a changes its sign, the complex monopole looks like a Gaussian beam with growing strength for decreasing negative values of a. This is shown in Fig. 2.2.

The close relationship of complex monopoles with Gaussian beams was described by Felsen [Fel76a], for example.

However, what happens for $(x, y, z) \in D$? In other words: What is the source for the complex monopole, i.e., the right-hand side of the Helmholtz equation

$$\Delta\varphi + k^2\varphi = -\tilde{\delta} \quad \text{for} \quad (x, y, z) \in \mathbb{R}^3? \tag{2.9}$$

for $\vec{r}_c = (x, y, z + ia)^T$, $\tilde{r} = \sqrt{x^2 + y^2 + (z + ia)^2}$ with a real constant a. Before we consider this question in Chap. 4, we will show that the complex monopole takes a very simple shape in oblate spheroidal coordinates.

Chapter 3
Complex Monopoles in Oblate Spheroidal Coordinates

Kaiser [Kai00] has mentioned that the oblate spheroidal coordinate system is closely related to the complex distance function in $\mathbb{R}^n$. We will now consider this relationship for the complex distance $\sqrt{x^2 + y^2 + (z + ia)^2}$ of the slightly simplified complex monopole (2.1)

$$\varphi(\vec{r}_c) = \varphi(x, y, z + ia) = \frac{e^{ik\sqrt{x^2+y^2+(z+ia)^2}}}{\sqrt{x^2 + y^2 + (z + ia)^2}} = \frac{e^{ik\tilde{r}}}{\tilde{r}}. \tag{3.1}$$

in three dimensions ($n = 3$). By using the definition of the oblate spheroidal coordinates (ζ, ξ, ϕ) (see [WikOs] and [Smy68]) with $0 \leq \zeta < \infty$ and $-1 \leq \xi \leq 1$

$$\begin{pmatrix} x \\ y \\ z \end{pmatrix} = a \begin{pmatrix} \sqrt{(1 + \zeta^2)(1 - \xi^2)} \cos\phi \\ \sqrt{(1 + \zeta^2)(1 - \xi^2)} \sin\phi \\ \zeta\xi \end{pmatrix}, \tag{3.2}$$

we immediately get for the real and imaginary parts p and q of the complex root

$$\tilde{r} = \sqrt{\alpha + i\beta} = \sqrt{x^2 + y^2 + (z + ia)^2} = p + iq = a(\zeta + i\xi). \tag{3.3}$$

On the other hand, the well-known formula for the real and imaginary parts of the complex root is

$$p = a\zeta = \frac{1}{\sqrt{2}}\sqrt{\sqrt{\alpha^2 + \beta^2} + \alpha}, q = a\xi = \frac{1}{\sqrt{2}}\sqrt{\sqrt{\alpha^2 + \beta^2} - \alpha \operatorname{sgn}(\beta)} \tag{3.4}$$

with $\alpha = x^2+y^2+z^2-a^2$, $\beta = 2az$. Thus, we have obtained a simple expression for the spheroidal coordinates ζ, ξ in terms of x, y, z. In these coordinates, the complex monopole (3.1) takes the form

M. Ochmann and R. Piscoya, *Theory and Application of Acoustic Sources Using Complex Analysis*, https://doi.org/10.1007/978-981-33-6040-2_3

$$\varphi(\vec{r}_c) = \tilde{\varphi}(\zeta, \xi, a) = \frac{e^{ika(\zeta+i\xi)}}{a(\zeta + i\xi)}. \tag{3.5}$$

Now, it can be shown—and much easier than in Cartesian coordinates—that φ fulfills the Helmholtz equation in oblate spheroidal coordinates for $w := \zeta + i\xi \neq 0$.

Proof We first have to determine the Laplacian operator in oblate spheroidal coordinates (see [WikOs]), which is given by

$$\begin{aligned}\Delta\varphi(\vec{r}_c) &= \Delta\tilde{\varphi}(\zeta, \xi, a)\\ &= \frac{1}{a^2(\zeta^2+\xi^2)}\left\{\frac{\partial}{\partial\zeta}\left[\left(1+\zeta^2\right)\frac{\partial\tilde{\varphi}}{\partial\zeta}\right] + \frac{\partial}{\partial\xi}\left[\left(1-\xi^2\right)\frac{\partial\tilde{\varphi}}{\partial\xi}\right]\right\},\end{aligned} \tag{3.6}$$

where we have taken into account that $\frac{\partial}{\partial\phi}\tilde{\varphi} = 0$. By using the complex number $w = \zeta + i\xi$, we get immediately (for simplicity, the tilde above φ is omitted)

$$\varphi = \frac{e^{ikaw}}{aw}, \varphi' = \frac{d\varphi}{dw} = \varphi\left(ika - \frac{1}{w}\right), \varphi'' = -\varphi\left(\frac{2}{w}\left(ika - \frac{1}{w}\right) + (ka)^2\right). \tag{3.7}$$

Since we have $\frac{\partial\varphi}{\partial\zeta} = \frac{d\varphi}{dw}\frac{\partial w}{\partial\zeta} = \varphi'$ and $\frac{\partial\varphi}{\partial\xi} = \frac{d\varphi}{dw}\frac{\partial w}{\partial\xi} = i\varphi'$, we obtain for the elements of the Laplacian

$$A := \frac{\partial}{\partial\zeta}\left[\left(1+\zeta^2\right)\frac{\partial\varphi}{\partial\zeta}\right] = 2\zeta\varphi' + \left(1+\zeta^2\right)\varphi'' \text{ and} \tag{3.8}$$

$$B := \frac{\partial}{\partial\xi}\left[\left(1-\xi^2\right)\frac{\partial\varphi}{\partial\xi}\right] = -2i\xi\varphi' - \left(1-\xi^2\right)\varphi''. \tag{3.9}$$

Summation of expressions A and B gives

$$A + B = -\varphi(ka)^2\left(\zeta^2+\xi^2\right), \tag{3.10}$$

and hence, we obtain the desired result

$$\Delta\varphi = \frac{1}{a^2\left(\zeta^2+\xi^2\right)}(A+B) = \frac{-\varphi(ka)^2\left(\zeta^2+\xi^2\right)}{a^2\left(\zeta^2+\xi^2\right)} = -k^2\varphi. \quad \square \tag{3.11}$$

The result that φ fulfils the homogeneous Helmholtz equation is only valid for $w := \zeta + i\xi \neq 0$ or equivalently for $(x, y, z) \notin D$. Thus, we come back to the question what happens for $(x, y, z) \in D$. In other words, what does the corresponding inhomogeneous Helmholtz equation look like?

Chapter 4
The Driving Source of the Complex Monopole

A first naive guess is the following: Since the "real monopole" with $a = 0$ is simple a solution of the inhomogeneous Helmholtz equation

$$\Delta\varphi + k^2\varphi = -\delta(x, y, z) \quad \text{for} \quad (x, y, z) \in \mathbb{R}^3 \tag{4.1}$$

with the usual Dirac delta function, we may suspect that the complex monopole is the solution of

$$\Delta\varphi + k^2\varphi = -\delta(x, y, z + ia) \quad \text{for} \quad (x, y, z) \in \mathbb{R}^3. \tag{4.2}$$

In fact, the Dirac delta function can be extended into the complex domain and is then called an ultradistribution (see, e.g., [Zem65], Chap. 7.7). However, it turns out that this assumption is wrong, which can be easily seen as it was communicated to one of the authors (Martin Ochmann) by Dr. Norbert Gorenflo.

The complex monopole φ is a discontinuous function and hence can be considered as a classical distribution. Thus, $\delta(x, y, z + ia) = -\left(\Delta\varphi + k^2\varphi\right)$ has to be a classical distribution, too (see Theorem 2.4-1 and below of [Zem65]). However, this is a contradiction, since the extended Dirac delta function is an ultradistribution ([Zem65], Chap. 7.7), but not a classical distribution.

The easiest way to find the correct source distribution is to consider first the one-dimensional case. It will turn out that this is an important key for the further construction of the theory in two and in three dimensions.

M. Ochmann and R. Piscoya, *Theory and Application of Acoustic Sources Using Complex Analysis*, https://doi.org/10.1007/978-981-33-6040-2_4

4.1 The Driving Source of the Complex Monopole in One Dimension (1D)

It is well known that the solution of the one-dimensional inhomogeneous Helmholtz equation

$$\frac{\mathrm{d}^2 y(z)}{\mathrm{d}z^2} + k^2 y(z) = -\delta(z - z_0) \tag{4.3}$$

is the 1D Green's function and is simply given by (see [Duf01], p. 275)

$$g_1(z) = y(z) = \frac{i}{2k}\mathrm{e}^{ik|z-z_0|}. \tag{4.4}$$

Here, we carefully have to distinguish between the monopole or Green's function or point source and the right-hand side (RHS) of the differential Eq. (4.3), which is called a source, too. Thus, the term "source" can lead to confusion, since it is applied on the one hand to the inhomogeneous part of Eq. (4.3) and on the other hand to the solution of Eq. (4.3) itself. Whenever such a misunderstanding may occur, we will call the RHS the driving source [in general, neglecting the minus sign appearing at the RHS of Eq. (4.3)]. For example, the driving source of the acoustical point source is the Dirac delta function.

The corresponding complex one-dimensional monopole (2.1) is obtained by choosing the imaginary source position $z_0 = -ia$:

$$y(z) = \frac{i}{2k}\mathrm{e}^{ik\sqrt{(z+ia)^2}}. \tag{4.5}$$

Just as in Eq. (4.4), we have to choose the sign of the complex square root in such a way that its real part is always positive, in order to guarantee that we obtain a radiating wave function due to the time convention $\exp(-i\omega t)$:

$$y(z) := \varphi(z + ia) = \frac{i}{2k}\begin{cases} \mathrm{e}^{ik(z+ia)} & \text{for } z > 0 \\ \mathrm{e}^{-ik(z+ia)} & \text{for } z < 0 \end{cases} \tag{4.6}$$

or

$$y(z) = \frac{i}{2k}\mathrm{e}^{ik|z|-ka\,\mathrm{sgn}(z)} \tag{4.7}$$

which gives the counterpart to the "real" one-dimensional monopole.

Thus, it can be observed that in addition to the discontinuous absolute value function now the even more singular signum function (with a jump at $z = 0$)

$$\operatorname{sgn}(z) = \begin{cases} 1 & \text{for } z > 0 \\ -1 & \text{for } z < 0 \end{cases} \tag{4.8}$$

occurs in the argument of $y(z)$ in Eq. (4.7). This complex 1D monopole fulfills the homogeneous 1D Helmholtz equation (4.3) for $z \neq 0$, which is simple a linear, ordinary second-order differential equation. In order to see what happens at $z = 0$, $y(z)$ given in Eq. (4.7) is inserted into the left-hand side of Eq. (4.3) with the following result:

Theorem 1 ***The driving source of the complex 1D monopole***
The function

$$g_1(z) = g_1(z, k) := y(z) = \frac{i}{2k} e^{ik\sqrt{(z+ia)^2}} = \frac{i}{2k} e^{ik|z| - ka\,\operatorname{sgn}(z)}, \tag{4.9}$$

where the square root has been chosen in such a way that its real part is always positive, is a solution of the inhomogeneous differential equation

$$\frac{d^2 y(z)}{dz^2} + k^2 y(z) = -\tilde{\delta}_1(z). \tag{4.10}$$

The right-hand side of Eq. (4.10) is the driving source of the complex 1D monopole given by

$$\tilde{\delta}_1(z) = \cosh(ka)\delta(z) + (i/k)\sinh(ka)\delta'(z). \tag{4.11}$$

Proof The proof can be conducted in three ways.

1. We observe that $y(z)$ can be written as

$$\begin{aligned} y(z) &= \frac{i}{2k} e^{ik|z| - ka\,\operatorname{sgn}(z)} = \frac{i}{2k} e^{ik|z|} e^{-ak\,\operatorname{sign}(z)} \\ &= \frac{i}{2k} e^{ik|z|} \{\cosh(ka\operatorname{sgn}(z)) - \sinh(ka\operatorname{sgn}(z))\} \\ &= \frac{i}{2k} e^{ik|z|} \{\cosh(ka) - \operatorname{sgn}(z)\sinh(ka)\} \\ &= \frac{i}{2k} \left\{ e^{ik|z|}\cosh(ka) + \left(e^{ik|z|}\right)' \frac{i}{k}\sinh(ka) \right\}, \end{aligned} \tag{4.12a}$$

since the derivative of $e^{ik|z|}$ is $\frac{d}{dz} e^{ik|z|} = \left(e^{ik|z|}\right)' = ike^{ik|z|}\operatorname{sgn}(z)$ and $\cosh(z)$ is an even and $\sinh(z)$ an odd function. On the other hand, it is well known that the one-dimensional Helmholtz operator $D_1 y := y'' + k^2 y$ gives $D_1\left(\frac{i}{2k} e^{ik|z|}\right) = -\delta(z)$ and by linearity $D\left(\frac{i}{2k} e^{ik|z|}\cosh(ka)\right) = -\cosh(ka)\delta(z)$, and finally

$$D_1\left(\frac{i}{2k}\left\{ e^{ik|z|}\cosh(ka) + \left(e^{ik|z|}\right)' \frac{i}{k}\sinh(ka) \right\}\right)$$

$$= -\left[\cosh(ka)\delta(z) + (i/k)\sinh(ka)\delta'(z)\right] = -\tilde{\delta}_1 \tag{4.12b}$$

which proves the claim (4.11).

2. The proof can also be conducted by using the theory of weak derivatives (see, e.g., [WikWD]), which is straightforward and is therefore not presented here.
3. In [Tag11], Eq. (11), a general rule for differentiating a function with jump discontinuities in generalized sense is used. By applying this method to $y(z)$, where the only jump occurs at $z = 0$, we obtain

$$D_1 y(z) = \left[y'(z)\right]_{z=0}\delta(z) + [y(z)]_{z=0}\delta'(z)$$

with jumps (see Eq. 4.6)

$$[y(z)]_{z=0} = y(z)\big|_{z\downarrow 0} - y(z)\big|_{z\uparrow 0} = \frac{i}{2k}\left(e^{-ka} - e^{ka}\right) = -\frac{i}{k}\sinh(ka)$$

and quite analogously

$$\left[y'(z)\right]_{z=0} = -\cosh(ka).$$

In summary, we get again

$$y'' + k^2 y = D_1 y = -\left[\cosh(ka)\delta(z) + (i/k)\sinh(ka)\delta'(z)\right] = -\tilde{\delta}_1.$$

□

Thus, in one dimension, the driving source of the complex monopole is not an ultradistribution, but a superposition of a Dirac delta function and its derivative. Due to the jump of the complex monopole $y(z)$, its source consists of a monopole contribution $M_1(z)\delta(z)$ and a dipole contribution $i D_1(z)\delta'(z)$

$$\tilde{\delta}_1 = \cosh(ka)\delta(z) + \frac{i}{k}\sinh(ka)\delta'(z) = M_1(z)\delta(z) + i D_1(z)\delta'(z). \tag{4.13}$$

The knowledge of the driving source in 1D can be used for deriving the 2D and the 3D driving sources. For that purpose, we first have to establish an extension of the Sommerfeld integral (see [Duf01], pp. 103 or the original work of Sommerfeld: [Som09], [Som26], [Som78, Chap. VI, Sect. 31]) to a complex position.

4.2 The Sommerfeld Integral for Real and Complex Source Locations

The well-known Sommerfeld integral (or the related Weyl integral if the plane wave decomposition is considered) is the representation of a monopole in cylindrical waves [see [Duf01], Eq. (3.5.29)]

$$\frac{e^{ik\sqrt{r^2+(z-z_0)^2}}}{4\pi\sqrt{r^2+(z-z_0)^2}} = \frac{i}{4\pi}\int\limits_0^\infty \frac{e^{i|z-z_0|\sqrt{k^2-k_r^2}}}{\sqrt{k^2-k_r^2}} J_0(k_r r)k_r \mathrm{d}k_r, \tag{4.14}$$

where J_0 is the Bessel functions of order 0 and $r^2 = x^2 + y^2$. The integral representation remains valid, if we choose an imaginary source position, for example $z_0 = -ia$:

$$\begin{aligned} g_3(x,y,z) &= \frac{e^{ik\sqrt{r^2+(z+ia)^2}}}{4\pi\sqrt{r^2+(z+ia)^2}} = \frac{i}{4\pi}\int\limits_0^\infty \frac{e^{i\sqrt{(z+ia)^2}\sqrt{k^2-k_r^2}}}{\sqrt{k^2-k_r^2}} J_0(k_r r)k_r \mathrm{d}k_r \\ &= \frac{i}{4\pi}\int\limits_0^\infty \frac{e^{i\,\mathrm{sgn}(z)(z+ia)\sqrt{k^2-k_r^2}}}{\sqrt{k^2-k_r^2}} J_0(k_r r)k_r \mathrm{d}k_r \end{aligned} \tag{4.15a}$$

where the branch $\sqrt{(z+ia)^2} = \mathrm{sgn}(z)(z+ia)$ has to be used for making the roots unique and for guaranteeing the convergence of the integral (see Eq. 4.6). Li and White [Li96] has performed a purely numerical investigation of the range of validity of Eq. (4.15a), since they remarked that "However, no one has proven that the Sommerfeld identity remains valid when the position becomes a complex number" ([Li96], p. 757). But, in the meantime, the proof was given by G. Kaiser in his topical review [Kai03] at p. 321, Theorem 4. Kaiser denotes it the "generalized Weyl formula" or "angular spectrum representation."

It is worth to note that due to the extended Sommerfeld integral, the complex 3D Green's function can be represented as the inverse Hankel transform of the complex 1D Green's function (4.9):

$$\begin{aligned} g_3 &= \frac{1}{4\pi}\frac{e^{ik\sqrt{r^2+(z+ia)^2}}}{\sqrt{r^2+(z+ia)^2}} = \frac{1}{2\pi}\int\limits_0^\infty \frac{i}{2}\frac{e^{i\sqrt{(z+ia)^2}\sqrt{k^2-k_r^2}}}{\sqrt{k^2-k_r^2}} J_0(k_r r)k_r \mathrm{d}k_r \\ &= \frac{1}{2\pi}H_r^{-1}(g_1(z,\tilde{\kappa})). \end{aligned} \tag{4.15b}$$

Here

$$H_r^{-1}(F) := \int_0^\infty F(k_r) J_0(k_r r) k_r \mathrm{d}k_r$$

denotes the inverse Hankel transform of order zero of the function F (see [Dav02], p. 227) and $\tilde{\kappa} = \sqrt{k^2 - k_r^2}$.

The 2D analogue of the Sommerfeld integral can be obtained by starting with the Green's function of the 2D Helmholtz equation and using an inverse 1D Fourier transform instead of the above inverse Hankel transform:

$$\begin{aligned} g_2(x,z) : &= \frac{i}{4} H_0^{(1)}\left(k\sqrt{x^2 + (z - z_0)^2}\right) = \frac{1}{2\pi} \int_{-\infty}^{\infty} \frac{i}{2\kappa} \mathrm{e}^{i\kappa|z - z_0|} \mathrm{e}^{ik_x x} \mathrm{d}k_x \\ &= F_x^{-1}(g_1(z,\kappa)), \end{aligned} \tag{4.16}$$

where $H_0^{(1)}$ denotes the Hankel function of the first kind and zero order and $\kappa = \sqrt{k^2 - k_x^2}$. Equation (4.16) shows that the 2D Green's function $g_2(x, z)$ (see, e.g., [Duf01] pp. 278) can be obtained by an inverse Fourier transform (denoted by F_x^{-1}) of the 1D Green's function $g_1(z, \kappa)$, see also Eq. (4.4). Equation (4.16) is a standard result and also known as the plane wave decomposition of the Hankel function. It can be proven in many different ways (see, e.g., [Duf01], p. 278, [Mor53], part I, p. 823 or [Gor94], pp. 790, Appendix). Again, we assume that the identity (4.16) can be extended to the situation with a complex source point $z_0 = -ia$ which leads to

$$\begin{aligned} g_2(x,z) &:= \frac{i}{4} H_0^{(1)}\left(k\sqrt{x^2 + (z + ia)^2}\right) = \frac{1}{2\pi} \int_{-\infty}^{\infty} \frac{i}{2\kappa} \mathrm{e}^{i\kappa\sqrt{(z+ia)^2}} \mathrm{e}^{ik_x x} \mathrm{d}k_x \\ &= \frac{1}{2\pi} \int_{-\infty}^{\infty} \frac{i}{2\kappa} \mathrm{e}^{i\kappa \operatorname{sgn}(z)(z+ia)} \mathrm{e}^{ik_x x} \mathrm{d}k_x = F_x^{-1}(g_1(z,\kappa)). \end{aligned} \tag{4.17}$$

The proof of Eq. (4.17) can be performed as follows (see also the analogous proof of [Kai03] for the Sommerfeld integral (4.15a) in 3D).

Proof of Eq. (4.17):

We use formula 3.9.14 (4) of [Gra07]—in edition [Gra80], this integral is missing—which can be written by substituting $x \to k_x, b \to x$ as

$$\begin{aligned} &\int_0^\infty \frac{\mathrm{e}^{-\beta\sqrt{\gamma^2 + k_x^2}}}{\sqrt{\gamma^2 + k_x^2}} \cos(k_x x) \mathrm{d}k_x = \frac{1}{2} \int_{-\infty}^{\infty} \frac{\mathrm{e}^{-\beta\sqrt{\gamma^2 + k_x^2}}}{\sqrt{\gamma^2 + k_x^2}} \exp(ik_x x) \mathrm{d}k_x \\ &= K_0\left(\gamma\sqrt{\beta^2 + x^2}\right), \quad \mathrm{Re}\beta > 0, \quad \mathrm{Re}\gamma > 0, \end{aligned} \tag{4.17a}$$

where K_0 is the modified Bessel function of second kind and order zero. Further, we assume that $\gamma = -ik + \varepsilon$, $\varepsilon \to 0$ with a tiny positive real part ε and a positive real wave number k for satisfying the condition $\mathrm{Re}\gamma > 0$. (Note that this is equivalent to suppose that the wave number becomes complex by $\tilde{k} = k + i\varepsilon$ with $k, \varepsilon > 0$, which assumption is used to proof the validity of the Sommerfeld–Weyl integral for real coordinates, see [Gor94], p. 790).

This leads to

$$\frac{1}{2}\int\limits_{-\infty}^{\infty} \frac{\mathrm{e}^{-\beta\sqrt{k_x^2-k^2}}}{\sqrt{k_x^2-k^2}} \exp(ik_x x)\mathrm{d}k_x = \frac{\pi i}{2} H_0^{(1)}\left(k\sqrt{\beta^2+x^2}\right), \tag{4.17b}$$

where we have used formula 9.6.4 on p. 375 of [Abr72].

Following Kaiser [Kai03], p. 320 by defining the branch

$$\mu = \sqrt{k_x^2 - k^2} = \begin{cases} +\sqrt{k_x^2-k^2}, & k_x^2 > k^2 \\ -i\sqrt{k^2-k_x^2}, & k_x^2 < k^2 \end{cases}, \tag{4.17c}$$

we obtain (symbolically written by inserting the lower definition for μ)

$$\int\limits_{-\infty}^{\infty} \frac{\mathrm{e}^{i\beta\sqrt{k^2-k_x^2}}}{\sqrt{k^2-k_x^2}} \exp(ik_x x)\mathrm{d}k_x = \pi H_0^{(1)}\left(k\sqrt{\beta^2+x^2}\right). \tag{4.17d}$$

Now, we choose $\beta = +\sqrt{(z+ia)^2} = z + ia$ and obtain for $\mathrm{Re}\beta = z > 0$

$$\int\limits_{-\infty}^{\infty} \frac{\mathrm{e}^{i(z+ia)\sqrt{k^2-k_x^2}}}{\sqrt{k^2-k_x^2}} \exp(ik_x x)\mathrm{d}k_x = \pi H_0^{(1)}\left(k\sqrt{(z+ia)^2+x^2}\right), \tag{4.17e}$$

which is Eq. (4.17). For $z < 0$, we choose

$$\beta = -\sqrt{(z+ia)^2} = -z - ia,$$

so that we have again $\mathrm{Re}\beta = -z > 0$ which leads to

$$\int\limits_{-\infty}^{\infty} \frac{\mathrm{e}^{-i(z+ia)\sqrt{k^2-k_x^2}}}{\sqrt{k^2-k_x^2}} \exp(ik_x x)\mathrm{d}k_x = \pi H_0^{(1)}\left(k\sqrt{(z+ia)^2+x^2}\right). \tag{4.17f}$$

Thus, for all real z we obtain

$$\int_{-\infty}^{\infty} \frac{e^{i(z+ia)\mathrm{sgn}(z)\sqrt{k^2-k_x^2}}}{\sqrt{k^2-k_x^2}} \exp(ik_x x)\mathrm{d}k_x = \pi H_0^{(1)}\left(k\sqrt{(z+ia)^2+x^2}\right). \tag{4.17g}$$

Obviously, the convergence of the integral can be seen better by looking at Eq. (4.17b) with

$$\beta = \sqrt{(z+ia)^2} = \mathrm{sgn}(z)(z+ia) = |z| + ia\mathrm{sgn}(z).$$

On the other hand, Formula (4.17g) shows that the integrand is the Fourier transform of the 1D complex monopole $g_1(z,\kappa)$ with respect to the wave number $\kappa = \sqrt{k^2 - k_x^2}$ given in Eq. (4.9).

□

4.3 The Driving Source of the Complex Monopole in Two Dimensions (2D)

We now consider the two-dimensional inhomogeneous Helmholtz equation

$$\frac{\partial^2 g_2(x,z)}{\partial x^2} + \frac{\partial^2 g_2(x,z)}{\partial z^2} + k^2 g_2(x,z) = -\delta(x)\delta(z-z_0), \tag{4.18}$$

where the solution is the 2D Green's function (4.16) (see [Duf01], p. 277)

$$g_2(x,z) = \frac{i}{4} H_0^{(1)}\left(k\sqrt{x^2+(z-z_0)^2}\right). \tag{4.19}$$

Just as in the 1D case, the corresponding complex two-dimensional monopole is obtained by choosing the imaginary source position $z_0 = -ia$:

$$g_2(x,z) = \frac{i}{4} H_0^{(1)}\left(k\sqrt{x^2+(z+ia)^2}\right). \tag{4.20}$$

Again, the root has to be chosen such that its real part is always positive to satisfy the radiation condition for the time convention $\exp(-i\omega t)$ taking into account the asymptotic expansion for the Hankel function of the first kind for large arguments (see [Abr72], p. 364, 9.2.3). Mathematically spoken, the function has to be made single-valued by introducing a suitable branch cut (see [Kai00], [Kai03], [Tag15], and our book Chap. 7). Tagirdzhanov et al. [Tag15] called this special cut the "source choice." For deriving the driving source $\tilde{\delta}_2(x,z)$ of the complex 2D monopole (4.20), Theorem 1 can be used together with the extended plane wave representation (4.17) of the Hankel function.

Theorem 2 ***The driving source of the complex 2D monopole***
The function

$$g_2(x,z) = \frac{i}{4} H_0^{(1)}\left(k\sqrt{x^2 + (z+ia)^2}\right), \tag{4.20}$$

where the square root has been chosen in such a way that the real part of $\sqrt{x^2 + (z+ia)^2}$ is always positive, is a solution of the inhomogeneous differential equation

$$\frac{\partial^2 g_2(x,z)}{\partial x^2} + \frac{\partial^2 g_2(x,z)}{\partial z^2} + k^2 g_2(x,z) = -\tilde{\delta}_2(x,z). \tag{4.21}$$

Here, the right-hand side of Eq. (4.21) is the driving source of the complex 2D monopole given by

$$\tilde{\delta}_2(x,z) = M_2\delta(z) + iD_2\delta'(z) \tag{4.22}$$

with

$$M_2 = \frac{1}{2\pi}\int_{-\infty}^{\infty} \cosh(\kappa a) e^{ik_x x} dk_x = \frac{ka}{2}\chi_{(-a,a)}(x)\frac{J_1\left(k\sqrt{x^2-a^2}\right)}{\sqrt{x^2-a^2}}$$
$$+ \frac{1}{2}\{\delta(x+a) + \delta(x-a)\} \tag{4.23}$$

and

$$D_2 = \frac{1}{2\pi}\int_{-\infty}^{\infty} \frac{1}{\kappa}\sinh(\kappa a) e^{ik_x x} dk_x = \frac{1}{2}\chi_{(-a,a)}(x) J_0\left(k\sqrt{x^2-a^2}\right), \tag{4.24}$$

where J_0 and J_1 are the Bessel functions of order 0 and 1, respectively, $\kappa = \sqrt{k^2 - k_x^2}$, and

$$\chi_{(-a,a)}(x) = \begin{cases} 1, & x \subset (-a,a) \\ 0 & \text{elsewhere} \end{cases} \tag{4.25}$$

denotes the characteristic function of the interval $(-a, a)$.

Proof: The Fourier transform of the 2D Helmholtz equation (4.21) with respect to x gives

$$\frac{\mathrm{d}^2 \hat{g}_2(z, k_x)}{\mathrm{d}z^2} + \kappa^2 \hat{g}_2(z, k_x) = -F_x\left(\tilde{\delta}_2(x, z)\right), \tag{4.26}$$

where $\hat{g}_2(z, k_x) := F_x(g_2(x, z))$ denotes the Fourier transform in x. Due to Eq. (4.17) $\hat{g}_2(z, k_x) = g_1(z, \kappa)$ and thus

$$\frac{\mathrm{d}^2 g_1(z, \kappa)}{\mathrm{d}z^2} + \kappa^2 g_1(z, \kappa) = -F_x\left(\tilde{\delta}_2(x, z)\right). \tag{4.27}$$

On the other hand, we know from Theorem 1 that the right-hand side of Eq. (4.27) is

$$F_x\left(\tilde{\delta}_2(x, z)\right) = \tilde{\delta}_1(z) = \cosh(\kappa a)\delta(z) + (i/\kappa)\sinh(\kappa a)\delta'(z) \tag{4.28}$$

where we only have to substitute the wave number k by $\kappa = \sqrt{k^2 - k_x^2}$. Thus, we get

$$\tilde{\delta}_2(x, z) = F_x^{-1}\left\{\cosh(\kappa a)\delta(z) + (i/\kappa)\sinh(\kappa a)\delta'(z)\right\}, \tag{4.29}$$

which can be written in a more concise presentation

$$\tilde{\delta}_2(x, z) = M_2\delta(z) + i D_2\delta'(z) \tag{4.30}$$

with

$$M_2 = \frac{1}{2\pi}\int_{-\infty}^{\infty} \cosh(\kappa a)\mathrm{e}^{ik_x x}\mathrm{d}k_x \tag{4.31}$$

and

$$D_2 = \frac{1}{2\pi}\int_{-\infty}^{\infty} \frac{1}{\kappa}\sinh(\kappa a)\mathrm{e}^{ik_x x}\mathrm{d}k_x. \tag{4.32}$$

For solving these Fourier integrals, we start with the dipole contribution D_2. We have to show the validity of Eq. (4.24)

$$\frac{1}{2\pi}\int_{-\infty}^{\infty} \frac{1}{\kappa}\sinh(\kappa a)\mathrm{e}^{ik_x x}\mathrm{d}k_x = \frac{1}{\pi}\int_{0}^{\infty} \frac{\sinh\left(a\sqrt{k^2 - k_x^2}\right)}{\sqrt{k^2 - k_x^2}}\cos(k_x x)\mathrm{d}k_x$$
$$= \frac{1}{2}\chi_{(-a,a)}(x) J_0\left(k\sqrt{x^2 - a^2}\right). \tag{4.33}$$

From [Erd54] p. 26, formula 30, we find that

$$\int_0^\infty \frac{\sin\left(a\sqrt{\tilde{k}^2+k_x^2}\right)}{\sqrt{\tilde{k}^2+k_x^2}}\cos(k_x x)\mathrm{d}k_x = \frac{\pi}{2}J_0\left(\tilde{k}\sqrt{a^2-x^2}\right)\chi_{(0,a)}(x) \text{ for } x>0$$

and

$$\int_0^\infty \frac{\sin\left(a\sqrt{\tilde{k}^2+k_x^2}\right)}{\sqrt{\tilde{k}^2+k_x^2}}\cos(k_x x)\mathrm{d}k_x = \frac{\pi}{2}J_0\left(\tilde{k}\sqrt{a^2-x^2}\right)\chi_{(-a,0)}(x) \text{ for } x<0$$

which can be written together as

$$\int_0^\infty \frac{\sin\left(a\sqrt{\tilde{k}^2+k_x^2}\right)}{\sqrt{\tilde{k}^2+k_x^2}}\cos(k_x x)\mathrm{d}k_x = \frac{\pi}{2}J_0\left(\tilde{k}\sqrt{a^2-x^2}\right)\chi_{(-a,a)}(x) \text{ for } -\infty<x<\infty. \tag{4.34}$$

Thus, we get the desired result (4.33) under the condition that Formula (4.34) remains valid for a pure imaginary quantity $\tilde{k}=ik$, since $\sin(ix)=i\sinh(x)$ and therefore

$$\frac{\sin\left(a\sqrt{-k^2+k_x^2}\right)}{\sqrt{-k^2+k_x^2}} = \frac{\sin\left((\pm i)a\sqrt{k^2-k_x^2}\right)}{(\pm i)\sqrt{k^2-k_x^2}} = \frac{\sinh\left(a\sqrt{k^2-k_x^2}\right)}{\sqrt{k^2-k_x^2}}.$$

Such a continuation is permitted, which can be seen by performing a Fourier Cosine transform of Eq. (4.34)

$$\frac{\sin\left(a\sqrt{\tilde{k}^2+k_x^2}\right)}{\sqrt{\tilde{k}^2+k_x^2}} = \int_0^a \frac{\pi}{2}J_0\left(\tilde{k}\sqrt{a^2-x^2}\right)\cos(k_x x)\mathrm{d}x. \tag{4.35}$$

The left-hand side of Eq. (4.35) is an entire function of the parameter $\tilde{k}$, the integral on the right-hand side exists, and thus, Eqs. (4.34) and (4.35) are valid for arbitrary $\tilde{k}$.

Now let us turn to the monopole contribution M_2.

Obviously, the integral does not converge, but we can easily split off the divergent part by

$$M_2 = \frac{1}{2\pi}\int_{-\infty}^{\infty} \cos\left(\sqrt{k_x^2-k^2}a\right)\mathrm{e}^{ik_x x}\mathrm{d}k_x$$

$$= \frac{1}{2\pi} \int_{-\infty}^{\infty} \left[\cos\left(\sqrt{k_x^2 - k^2} a \right) - \cos(k_x a) \right] e^{ik_x x} dk_x$$

$$+ \frac{1}{2\pi} \int_{-\infty}^{\infty} \cos(k_x a) e^{ik_x x} dk_x =: I_1 + I_2. \tag{4.36}$$

The second integral I_2 gives

$$I_2 = \frac{1}{2\pi} \int_{-\infty}^{\infty} \cos(k_x a) e^{ik_x x} dk_x = \frac{1}{2} [\delta(x + a) + \delta(x - a)]. \tag{4.37}$$

The first integral I_1 is more difficult to solve. We have to show that

$$I_1 = \frac{ka}{2} \chi_{(-a,a)}(x) \frac{J_1\left(k\sqrt{x^2 - a^2}\right)}{\sqrt{x^2 - a^2}} \tag{4.38}$$

or equivalently by taking the Fourier Cosine transform of Eq. (4.38) that

$$\cos\left(\sqrt{k_x^2 - k^2} a \right) - \cos(k_x a) = ka \int_0^a \frac{J_1\left(ik\sqrt{a^2 - x^2}\right)}{i\sqrt{a^2 - x^2}} \cos(k_x x) dx. \tag{4.39}$$

We start with the integral on the RHS of assertion (4.39) by using the substitution

$$x = a \cos \vartheta, \quad \sqrt{a^2 - x^2} = a \sin \vartheta, \quad \frac{dx}{d\vartheta} = -a \sin \vartheta \tag{4.40}$$

which leads us to

$$\text{RHS}(4.39) = -ika \int_0^{\pi/2} J_1(ika \sin \vartheta) \cos(k_x a \cos \vartheta) d\vartheta =: kaJ. \tag{4.41}$$

A related integral can be found in [Gra80, p. 742, formula 6.688,1]:

$$\int_0^{\pi/2} J_\nu(\mu z \sin t) \cos(\mu x \cos t) dt$$

$$= \frac{\pi}{2} J_{\nu/2}\left(\mu \frac{\sqrt{x^2 + z^2} + x}{2} \right) J_{\nu/2}\left(\mu \frac{\sqrt{x^2 + z^2} - x}{2} \right) \tag{4.42}$$

which is valid for $\mathrm{Re}\nu > -1$, $\mathrm{Re}z > 0$. By choosing in (4.42)

$$\mu = a, \quad t = \vartheta, \quad x = k_x, \quad z = ik, \quad \nu = 1, \tag{4.43}$$

we obtain [cp. definition of J in (4.41)]

$$\begin{aligned} J &:= -i \int_0^{\pi/2} J_1(ika \sin\vartheta)\cos(k_x a\cos\vartheta)\mathrm{d}\vartheta \\ &= -i\frac{\pi}{2} J_{1/2}\left(a\frac{\sqrt{k_x^2 - k^2} + k_x}{2}\right) J_{1/2}\left(a\frac{\sqrt{k_x^2 - k^2} - k_x}{2}\right). \end{aligned} \tag{4.44}$$

The choice $z = ik$ is allowed, since the RHS of (4.42) is an entire function of z for $\nu = 1$ and the integral (on the LHS) exists. Thus, also the LHS of (4.42) is holomorph everywhere, and formula (4.42) is valid for all complex numbers z for $\nu = 1$. We owe this reasoning of the validity of Eqs. (4.42)/(4.44) to Norbert Gorenflo. Now, it is well known (see [Abr72], p. 437 and 438) that $J_{1/2}$ is related to the spherical Bessel function of order zero j_0 such that

$$J_{1/2}(z) = \sqrt{\frac{z}{\pi/2}}\, j_0(z) = \sqrt{\frac{2}{\pi}}\frac{\sin z}{\sqrt{z}}. \tag{4.45}$$

By inserting Eq. (4.45) into the RHS of (4.44), a short calculation gives

$$J := \frac{1}{ka}\left\{\cos\left(\sqrt{k_x^2 - k^2 a}\right) - \cos k_x a\right\}. \tag{4.46}$$

Comparing the result (4.46) with (4.41), we see that

$$\mathrm{RHS}(4.39) = \cos\left(\sqrt{k_x^2 - k^2 a}\right) - \cos k_x a = \mathrm{LHS}(4.39), \tag{4.47}$$

which completes the proof.

□

The result of Theorem 2 was also found by Tagirdzhanov in [Tag13] proceeding in a different way by using a direct calculation, Green's formula, jump relations, and a special regularization technique. By comparing formula (10) of [Tag13]—which he called the source choice—with Eq. (4.22) in Theorem 2, we see a complete agreement except that only the dipole contribution has a different sign. The reason is that we use the root $\sqrt{x^2 + (z + ia)^2}$ and Tagirdzhanov $\sqrt{x^2 + (z - ia)^2}$. The last choice leads to a minus sign in the dipole contribution only, so that we would obtain

$$\tilde{\delta}_2(x, z) = F_x^{-1}\left\{\cosh(\kappa a)\delta(z) - (i/\kappa)\sinh(\kappa a)\delta'(z)\right\}$$

instead of Eq. (4.29). This can be seen easily from Theorem 1 by changing a to $-a$ and taking into account that cosh is an even and sinh an odd function.

4.4 The Driving Source of the Complex Monopole in Three Dimensions (3D)

Now, we come back to the three-dimensional inhomogeneous Helmholtz equation, which clearly has the highest importance for applications:

$$\Delta g_3 + k^2 g_3 = -\tilde{\delta}_3. \tag{4.48}$$

We assume that the solution of Eq. (4.48) is the 3D complexified Green's function or the complex monopole

$$g_3 = \frac{1}{4\pi}\varphi(\vec{r}_c) = \frac{\mathrm{e}^{ik\sqrt{x^2+y^2+(z+ia)^2}}}{4\pi\sqrt{x^2+y^2+(z+ia)^2}} = \frac{\mathrm{e}^{ik\tilde{r}}}{4\pi\tilde{r}}. \tag{4.49}$$

Again, the root has to be chosen such that its real part is always positive as described in Chap. 2. For deriving the driving source $\tilde{\delta}_3$ of the complex 3D monopole, the 1D presentation of Theorem 1 can be used, together with the generalized Sommerfeld identity (4.15). By defining $r = \sqrt{x^2+y^2}$ and the Heaviside unit step function

$$H(x) = \begin{cases} 1, & x > 0 \\ 0, & x < 0 \end{cases} \tag{4.50}$$

we obtain:

Theorem 3 ***The driving source of the complex 3D monopole***
The function

$$g_3(x, y, z) = \frac{1}{4\pi}\frac{\mathrm{e}^{ik\sqrt{x^2+y^2+(z+ia)^2}}}{\sqrt{x^2+y^2+(z+ia)^2}} \tag{4.49}$$

where the branch of the root has been chosen in such a way that the real part of $\sqrt{x^2+y^2+(z+ia)^2}$ is always positive, is a solution of the inhomogeneous Helmholtz equation

$$\Delta g_3 + k^2 g_3 = -\tilde{\delta}_3(x, y, z). \tag{4.48}$$

Here, the right-hand side of Eq. (4.48) is the driving source of the complex 3D monopole given by

$$\tilde{\delta}_3(x, y, z) = M_3(r, a)\delta(z) + i D_3(r, a)\delta'(z), \tag{4.51}$$

where D_3 is the dipole contribution

$$\begin{aligned} D_3(r, a) &= \frac{1}{2\pi} \int_0^\infty \frac{1}{\kappa} \sinh(\kappa a) J_0(k_r r) k_r \mathrm{d}k_r \\ &= \frac{H(a - r)}{2\pi} \frac{\cosh\left(k\sqrt{a^2 - r^2}\right)}{\sqrt{a^2 - r^2}}. \end{aligned} \tag{4.52}$$

The monopole contribution is—just as in the 2D case—not an ordinary function, but a distribution symbolically written as

$$\begin{aligned} M_3 &= \frac{1}{2\pi} \int_0^\infty \cosh(\kappa a) J_0(k_r r) k_r \mathrm{d}k_r \\ &= \frac{1}{2\pi} \left\{ \frac{\delta(r - a)}{r} \cosh(ka) - a H(a - r) \right. \\ &\left. \left[\frac{\cosh\left(k\sqrt{a^2 - r^2}\right)}{\left(a^2 - r^2\right)^{3/2}} - k \frac{\sinh\left(k\sqrt{a^2 - r^2}\right)}{a^2 - r^2} \right] [((\bullet)) - (a)] \right\} \end{aligned} \tag{4.53}$$

where the notation (4.53) means explicitly that the functional $M_3 = M_3(r, a)$ acts on sufficiently smooth test functions $f(r)$ such that

$$\begin{aligned} \langle M_3(r, a), f(r) \rangle = \frac{1}{2\pi} \left\{ f(a) \cosh(ka) - a \int_0^a \left[\frac{\cosh\left(k\sqrt{a^2 - r^2}\right)}{\left(a^2 - r^2\right)^{3/2}} \right. \right. \\ \left. \left. -k \frac{\sinh\left(k\sqrt{a^2 - r^2}\right)}{a^2 - r^2} \right] [f(r) - f(a)] r \mathrm{d}r \right\} \end{aligned} \tag{4.54}$$

Thus, $\langle M_3, f \rangle$ denotes the number which M_3 assigns to the test function f (see [Zem65], p. 6).

Proof: The Hankel transform of order zero (see, e.g., Chap. 14 of the text book of Davies [Dav02]) applied to the 3D Helmholtz equation (4.48) with respect to $r = \sqrt{x^2 + y^2}$ gives

$$\frac{d^2\hat{g}_3(z,k_r)}{dz^2}+\kappa^2\hat{g}_3(z)=-\int_0^\infty \tilde{\delta}_3 J_0(k_r r)r\mathrm{d}r \tag{4.55}$$

where $\hat{g}_3(z,k_r) := \int_0^\infty g_3(r,z)J_0(k_r r)r\mathrm{d}r$ denotes the Hankel transform H_r with respect to r, and $\kappa=\sqrt{k^2-k_r^2}$. Due to the extended Sommerfeld integral (4.15), g_3 can be written as the inverse Hankel transform of $g_1(z,\kappa)$ (see Eq. 4.15b)

$$\begin{aligned} g_3 &= \frac{1}{4\pi}\frac{\mathrm{e}^{ik\sqrt{r^2+(z+ia)^2}}}{\sqrt{r^2+(z+ia)^2}} = \frac{1}{2\pi}\int_0^\infty \frac{i}{2}\frac{\mathrm{e}^{i\sqrt{(z+ia)^2}\sqrt{k^2-k_r^2}}}{\sqrt{k^2-k_r^2}}J_0(k_r r)k_r\mathrm{d}k_r \\ &= \frac{1}{2\pi}H_r^{-1}(g_1). \end{aligned} \tag{4.56}$$

Thus, $\hat{g}_3(z,k_r)=\frac{1}{2\pi}g_1(z,\kappa)$ and Eq. (4.55) becomes

$$\frac{\mathrm{d}^2 g_1(z,\kappa)}{\mathrm{d}z^2}+\kappa^2 g_1(z,\kappa)=-2\pi H_r\Big(\tilde{\delta}_3(x,z)\Big). \tag{4.57}$$

On the other hand, we know from Theorem 1 that the right-hand side of Eq. (4.57) must be $-\big\{\cosh(\kappa a)\delta(z)+(i/\kappa)\sinh(\kappa a)\delta'(z)\big\}$ too, and thus we get

$$2\pi H_r\Big(\tilde{\delta}_3(x,z)\Big)=\tilde{\delta}_1(z)=\big\{\cosh(\kappa a)\delta(z)+(i/\kappa)\sinh(\kappa a)\delta'(z)\big\}, \tag{4.58}$$

where we only have to substitute the wave number k (appearing in Theorem 1) by $\kappa=\sqrt{k^2-k_r^2}$. Thus, we get

$$\tilde{\delta}_3(x,z)=\frac{1}{2\pi}H_r^{-1}\big\{\cosh(\kappa a)\delta(z)+(i/\kappa)\sinh(\kappa a)\delta'(z)\big\} \tag{4.59}$$

which can be written in the more concise presentation

$$\tilde{\delta}_3(x,z)=M_3\delta(z)+iD_3\delta'(z) \tag{4.60}$$

with

$$M_3=\frac{1}{2\pi}\int_0^\infty \cosh(\kappa a)J_0(k_r r)k_r\mathrm{d}k_r \tag{4.61}$$

and

$$D_3=\frac{1}{2\pi}\int_0^\infty \frac{1}{\kappa}\sinh(\kappa a)J_0(k_r r)k_r\mathrm{d}k_r. \tag{4.62}$$

By comparing Eqs. (4.61) and (4.62) with Eqs. (4.31) and (4.32), we see that the only difference between the 2D and the 3D case is that we have to use the Hankel transform instead of the Fourier transform. This makes sense, since the Hankel transform of order zero is a 2D Fourier transform of a circularly symmetric function (see [WikHa]), and the complex monopole g_3 is independent of the polar angle ϕ when considering a cylindrical coordinate system.

For solving these "Hankel integrals," we proceed similar as in 2D by starting with ***the dipole contribution*** D_3:

We have to show that Eq. (4.52) is valid:

$$\frac{1}{2\pi}\int_0^\infty \frac{1}{\kappa}\sinh(\kappa a)J_0(k_r r)k_r \mathrm{d}k_r = \frac{H(a-r)}{2\pi}\frac{\cosh\left(k\sqrt{a^2-r^2}\right)}{\sqrt{a^2-r^2}} \tag{4.63}$$

or

$$\int_0^\infty \frac{\sin\left(a\sqrt{k_r^2-k^2}\right)}{\sqrt{k_r^2-k^2}} J_0(k_r r)k_r \mathrm{d}k_r = H(a-r)\frac{\cosh\left(k\sqrt{a^2-r^2}\right)}{\sqrt{a^2-r^2}}. \tag{4.64}$$

This relationship can be proven by using formula 6.737,5 in [Gra80], p. 761. Alternatively, we perform the inverse Hankel transform of (4.64) leading to

$$\frac{\sin\left(a\sqrt{k_r^2-k^2}\right)}{\sqrt{k_r^2-k^2}} = \int_0^a \frac{\cosh\left(k\sqrt{a^2-r^2}\right)}{\sqrt{a^2-r^2}} J_0(k_r r) r \mathrm{d}r =: J. \tag{4.65}$$

By using the substitution

$$r = a\sin\tau, \quad \frac{\mathrm{d}r}{\mathrm{d}\tau} = a\cos\tau,$$

we obtain for the right-hand side of Eq. (4.65)

$$J = \int_0^{\pi/2} \cosh(ka\cos\tau)J_0(k_r a\sin\tau)a\sin\tau \mathrm{d}\tau. \tag{4.66}$$

Now, we can apply formula 6.6882 of [Gra80], p. 743, which gives for $\nu = 0$

$$\int_0^{\pi/2} \cos(\beta\cos\tau)J_0(\alpha\sin\tau)\sin\tau \mathrm{d}\tau = \sqrt{\frac{\pi}{2}}\frac{1}{\left(\alpha^2+\beta^2\right)^{\frac{1}{4}}}J_{1/2}\left(\sqrt{\alpha^2+\beta^2}\right)$$

$$= \frac{\sin\left(\sqrt{\alpha^2+\beta^2}\right)}{\sqrt{\alpha^2+\beta^2}}, \tag{4.67}$$

where we have used Formula (4.45) for the Bessel function of order 1/2.

If we choose for the parameters $\alpha = k_r a$ and $\beta = ika$, we obtain from Eqs. (4.66) and (4.67)

$$J = \int_0^{\pi/2} \cosh(ka\cos\tau) J_0(k_r a \sin\tau) a \sin\tau \mathrm{d}\tau = \frac{\sin\left(a\sqrt{k_r^2-k^2}\right)}{\sqrt{k_r^2-k^2}}, \tag{4.68}$$

which is the LHS of Eq. (4.65). This completes the proof of Eq. (4.65). Since J is an entire function of the parameter β, it is allowed to choose $\beta = ika$ in formula (4.67).

Now let us turn to the monopole contribution M_3:

We have to prove Formula (4.53) or equivalently its Hankel transform

$$\begin{aligned}
&\cosh\left(a\sqrt{k^2-k_r^2}\right) - J_0(k_r a)\cosh(ka) \\
&= a\int_0^a \left[-\frac{\cosh\left(k\sqrt{a^2-r^2}\right)}{\left(a^2-r^2\right)^{3/2}} + k\frac{\sinh\left(k\sqrt{a^2-r^2}\right)}{a^2-r^2}\right] \\
&\cdot [J_0(k_r r) - J_0(k_r a)] r\mathrm{d}r =: \mu
\end{aligned} \tag{4.69}$$

where we have used that the Hankel transform of $\delta(r)/r$ is given by

$$\int_0^\infty \frac{\delta(r-a)}{r} J_0(k_r r) r\mathrm{d}r = J_0(k_r a). \tag{4.70}$$

Clearly, the integral M_3 (4.61) is divergent due to appearance of the Dirac Delta function in the result (4.53), but the integral [which is the inverse Hankel transform of the left-hand side of Eq. (4.69)]

$$M_3^{\mathrm{reg}} = \int_0^\infty \left\{\cosh\left(a\sqrt{k^2-k_r^2}\right) - J_0(k_r a)\cosh(ka)\right\} J_0(k_r r) k_r \mathrm{d}r \tag{4.71}$$

is regular.

We now start with the right-hand side μ of Eq. (4.69). The first step is again the substitution

$$r = a \sin \tau, \quad \sqrt{a^2 - r^2} = a \cos \tau, \quad \frac{\mathrm{d}r}{\mathrm{d}\tau} = a \cos \tau \tag{4.72}$$

leading to

$$\mu = (ka)^2 \int_0^{\pi/2} \left[-\frac{\cosh(ka \cos \tau)}{(ka \cos \tau)^2} + \frac{\sinh(ka \cos \tau)}{ka \cos \tau} \right] \cdot [J_0(k_r a \sin \tau) - J_0(k_r a)] \sin \tau \mathrm{d}\tau. \tag{4.73}$$

By taking into account that

$$\frac{\mathrm{d}}{\mathrm{d}\tau}\left[\frac{\cosh(ka \cos \tau)}{ka \cos \tau}\right] = \left[\frac{\sinh(ka \cos \tau)}{ka \cos \tau} - \frac{\cosh(ka \cos \tau)}{(ka \cos \tau)^2}\right](-ka \sin \tau), \tag{4.74}$$

a partial integration can be performed as a second step

$$\begin{aligned}
\mu &= -(ka) \int_0^{\pi/2} \frac{\mathrm{d}}{\mathrm{d}\tau}\left[\frac{\cosh(ka \cos \tau)}{ka \cos \tau}\right][J_0(k_r a \sin \tau) - J_0(k_r a)]\mathrm{d}\tau \\
&= -(ka)\left\{\left[\frac{\cosh(ka \cos \tau)}{ka \cos \tau}\right][J_0(k_r a \sin \tau) - J_0(k_r a)]\right\}\Big|_0^{\pi/2} \\
&\quad + ka \int_0^{\pi/2} \frac{\cosh(ka \cos \tau)}{ka \cos \tau} \frac{\mathrm{d}}{\mathrm{d}\tau} J_0(k_r a \sin \tau)\mathrm{d}\tau =: \mu_1 + \mu_2.
\end{aligned} \tag{4.75}$$

The first term can be evaluated by using L'Hospital's rule: The upper limit at $\pi/2$ gives zero and the lower limit

$$\mu_1 = \cosh(ka)\{1 - J_0(k_r a)\}. \tag{4.76}$$

By using the well-known relationship (see [Abr72], p. 361, formula 9.1.28)

$$\frac{\mathrm{d}}{\mathrm{d}\tau} J_0(k_r a \sin \tau) = -J_1(k_r a \sin \tau) k_r a \cos \tau \tag{4.77}$$

the second term becomes

$$\mu_2 = -k_r a \int_0^{\pi/2} \cosh(ka \cos \tau) J_1(k_r a \sin \tau)\mathrm{d}\tau. \tag{4.78}$$

Now, we proceed along the same lines as in the 2D case. Formula 6.688,1 on p. 742 of [Gra80] with $\mu = a, z = k_r, x = ik, \nu = 1$ gives

$$\int_0^{\pi/2} J_1(k_r a \sin\tau)\cosh(ka\cos\tau)\mathrm{d}\tau = \frac{\pi}{2} J_{1/2}\left(a\frac{\sqrt{k_r^2 - k^2} + ik}{2}\right) \cdot J_{1/2}\left(a\frac{\sqrt{k_r^2 - k^2} - ik}{2}\right).$$

Thus, we obtain

$$\begin{aligned}\mu_2 &= -k_r a \int_0^{\pi/2} J_1(k_r a \sin\tau)\cosh(ka\cos\tau)\mathrm{d}\tau \\ &= \frac{-k_r a\pi}{2} J_{1/2}\left(a\frac{\sqrt{k_r^2 - k^2} + ik}{2}\right) J_{1/2}\left(a\frac{\sqrt{k_r^2 - k^2} - ik}{2}\right)\end{aligned}$$

or by taking into account Eqs. (4.45) and (4.46)

$$\begin{aligned}\mu_2 &= \frac{-k_r a\pi}{2} J_{1/2}\left(a\frac{\sqrt{k_r^2 - k^2} + ik}{2}\right) J_{1/2}\left(a\frac{\sqrt{k_r^2 - k^2} - ik}{2}\right) \\ &= \cosh\left(a\sqrt{k^2 - k_r^2}\right) - \cosh(ka).\end{aligned} \tag{4.79}$$

Thus, we finally obtain by adding Eqs. (4.76) and (4.79)

$$\mu = \mu_1 + \mu_2 = \cosh\left(a\sqrt{k^2 - k_r^2}\right) - J_0(k_r a)\cosh(ka), \tag{4.80}$$

which is the left-hand side of the assertion (4.69) and thus, the proof is complete.

□

Similar results as given in Theorem 3 were found earlier. If we let the wave number k go to zero, the potential case is obtained, and the driving sources (4.51) with $k = 0$ of Theorem 3 were found by G. Kaiser within the framework of complex-distance potential theory (see [Kai00], Proposition 3).

The surprisingly simple structure of the 2D and 3D complex monopole in Fourier space stems from the fact that they can be considered as inverse Fourier/Hankel transforms of the 1D complex monopole given in Theorem 1. This was also observed by Kaiser [Kai03] in his topical review in Corollary 2 [see his formula (118)], which leads him to the astonished expression: "The simplicity of the above expressions may be a 'miracle', but it cannot be an accident" (p. R317).

Tagirdzhanov et al. had given similar expressions for the source of a complex 3D monopole, again by employing Green's formula, jump relations, and a special regularization technique just as in 2D. These results can be found first in [Tag11]. However, the result for the source case, which we are considering, was not correct and was corrected in [Tag15]. On first sight, the expressions given by Tagirdzhanov et al. and here in Theorem 3 are different, but it can be shown that they are (nearly) equivalent in distributional sense with the exception of the pre-factor of the Delta function in Eq. (4.53). The proof is given in Appendix A.

Chapter 5
Application of Complex Sources for Half-Space Green's Functions Above an Impedance Plane

As mentioned in the introduction for motivation, complex sources are building blocks for constructing Green's functions in half-spaces, which allows to determine the sound field caused by monopoles over impedance planes. In [Och04], it was shown that the superposition integral of equivalent sources along a line in the mirror space below the impedance plane does only converge for specially chosen ground impedances. However, if these image sources are located at complex source points, the superposition integral converges for all kind of ground impedances. Now, this result will be generalized by formulating a "*general impedance formula.*"

5.1 General Impedance Formula

We formulate the general formula in the following theorem:

Theorem 4 ***General impedance formula (superposition integral)***
Let $u_0(z) = u_0(z, x, y, \ldots)$be a "basis" solution of a linear differential equation depending on $z \in (-\infty, \infty)$ and possibly also on other coordinates, symbolically written as

$$W(u) = 0 \tag{5.1a}$$

or

$$W(u) = -\delta(z - h), \quad h > 0. \tag{5.1b}$$

M. Ochmann and R. Piscoya, *Theory and Application of Acoustic Sources Using Complex Analysis*, https://doi.org/10.1007/978-981-33-6040-2_5

Then, the function

$$\hat{u}(z) = u_0(z) + u_0(-z) + 2\gamma \int_0^\infty e^{\gamma\tau} u_0(-z-\tau)\mathrm{d}\tau \tag{IF1}$$

or alternatively

$$\hat{u}(z) = u_0(z) + u_0(-z) + 2i\gamma \int_0^\infty e^{i\gamma\tau} u_0(-z-i\tau)\mathrm{d}\tau \tag{IF2}$$

is a solution of the field Eqs. (5.1a) or (5.1b) in the upper half-space V_+ with $z \in (0, \infty)$ satisfying the impedance or Robin or mixed boundary condition at the plane between both half-spaces at $z = 0$

$$\frac{\partial \hat{u}}{\partial z} + \gamma\hat{u} = 0 \quad \text{for } z = 0 \tag{IBC}$$

under two assumptions:

A_1 The integrals appearing in (IF1) and (IF2) have to be convergent.
A_2 The basis solution $u_0(z)$ is translational invariant in the following sense: $u_0(-z-\tau)$ or $u_0(-z-i\tau)$ is a solution of Eqs. (5.1a) or (5.1b) for all $\tau \in (0, \infty)$, too.

The line integral in (IF1) can be considered as a superposition of sources with real source positions (shortly: real representation). The line integral in (IF2) can be considered as a superposition of sources with complex source positions (shortly: complex representation).

Proof

(a) $\hat{u}$ clearly is a solution of the homogeneous field Eq. (5.1a) in V_+ due to its linearity, since it is a superposition of single solutions. Considering Eq. (5.1b), it is a solution of $W(u) = -\delta(z-h)$ in V_+, since all other singularities—except the singularity at $z = h$—produced by the Dirac delta function are located outside of the physical half-space $V_+ = \{z : z \in [0, \infty)\}$. This shows that the theorem remains valid, if the RHS of the field Eq. (5.1b) consists of concentrated sources with singularities located only in V_+. Such singularities will be mirrored by the impedance formula (IF1) into the image space V_- or by formula (IF2) into the complex domain.
(b) A direct calculation shows that formula (IF1) fulfills the boundary condition (IBC). We first consider the real representation (IF1):
By assuming that the order of integration and differentiation can be changed, differentiation of Eq. (IF1) leads to (the apostrophe $'$ denotes differentiation with respect to the argument)

$$\frac{\partial \hat{u}}{\partial z} = u_0'(z) - u_0'(-z) + 2\gamma \int_0^\infty e^{\gamma\tau} \frac{\partial}{\partial z} u_0(-z-\tau)d\tau = u_0'(z) - u_0'(-z)$$
$$+ 2\gamma \int_0^\infty e^{\gamma\tau} \frac{\partial}{\partial \tau} u_0(-z-\tau)d\tau$$

and integration by parts gives

$$\frac{\partial \hat{u}}{\partial z} = u_0'(z) - u_0'(-z) + 2\gamma e^{\gamma\tau} u_0(-z-\tau)|_0^\infty - 2\gamma \int_0^\infty \gamma e^{\gamma\tau} u_0(-z-\tau)d\tau.$$

It follows that

$$\frac{\partial \hat{u}}{\partial z} = u_0'(z) - u_0'(-z) - 2\gamma u_0(-z) - 2\gamma \int_0^\infty \gamma e^{\gamma\tau} u_0(-z-\tau)d\tau, \tag{5.2a}$$

where we have assumed that the contribution from the upper limit vanishes

$$e^{\gamma\tau} u_0(-z-\tau) \underset{\tau\to\infty}{\longrightarrow} 0. \tag{5.3a}$$

Thus we get

$$\frac{\partial \hat{u}}{\partial z}(z=0) = -2\gamma \left\{ u_0(0) + \int_0^\infty \gamma e^{\gamma\tau} u_0(-\tau)d\tau \right\}$$

or

$$\frac{\partial \hat{u}}{\partial z}(z=0) + \gamma \hat{u}(z=0) = -2\gamma \left\{ u_0(0) + \int_0^\infty \gamma e^{\gamma\tau} u_0(-\tau)d\tau \right\}$$
$$+ \gamma \left\{ 2u_0(0) + 2\gamma \int_0^\infty e^{\gamma\tau} u_0(-\tau)d\tau \right\} = 0,$$

which shows that the boundary condition (IBC) is fulfilled.
The proof for the complex representation (IF2) is analog, but should be given here, in order to derive the counterpart corresponding to condition (5.3a).
We have to differentiate Eq. (IF2) which leads to (here: $\partial/\partial z = \partial/\partial(i\tau)$)

$$\frac{\partial \hat{u}}{\partial z} = u_0'(z) - u_0'(-z) + 2i\gamma \int_0^\infty e^{i\gamma\tau} \frac{\partial}{\partial z} u_0(-z - i\tau)\mathrm{d}\tau$$
$$= u_0'(z) - u_0'(-z) + 2\gamma \int_0^\infty e^{i\gamma\tau} \frac{\partial}{\partial \tau} u_0(-z - i\tau)\mathrm{d}\tau.$$

Integration by parts gives

$$\frac{\partial \hat{u}}{\partial z} = u_0'(z) - u_0'(-z) + 2\gamma e^{\gamma\tau} u_0(-z - i\tau)|_0^\infty - 2\gamma \int_0^\infty i\gamma e^{i\gamma\tau} u_0(-z - i\tau)\mathrm{d}\tau.$$

It follows that

$$\frac{\partial \hat{u}}{\partial z} = u_0'(z) - u_0'(-z) - 2\gamma u_0(-z) - 2\gamma \int_0^\infty i\gamma e^{i\gamma\tau} u_0(-z - i\tau)\mathrm{d}\tau, \quad (5.2b)$$

where we now have to assume that

$$e^{i\gamma\tau} u_0(-z - i\tau) \underset{\tau\to\infty}{\longrightarrow} 0. \quad (5.3b)$$

Thus we get

$$\frac{\partial \hat{u}}{\partial z}(z = 0) = -2\gamma \left\{ u_0(0) + \int_0^\infty i\gamma e^{i\gamma\tau} u_0(-i\tau)\mathrm{d}\tau \right\}$$

or

$$\frac{\partial \hat{u}}{\partial z}(z = 0) + \gamma \hat{u}(z = 0) = -2\gamma \left\{ u_0(0) + \int_0^\infty i\gamma e^{\gamma\tau} u_0(-i\tau)\mathrm{d}\tau \right\}$$
$$+\gamma \left\{ 2u_0(0) + 2i\gamma \int_0^\infty e^{i\gamma\tau} u_0(-i\tau)\mathrm{d}\tau \right\} = 0$$

□

Formulas (IF1) and (IF2) can be generalized by choosing for the path of integration half rays from the origin to infinity in the complex plane $\{\zeta\tau = e^{i\psi}\tau, \tau \in [0, \infty)\}$ leading to

$$\hat{u}(z) = u_0(z) + u_0(-z) + 2\zeta\gamma \int_0^\infty e^{\zeta\gamma\tau} u_0(-z - \zeta\tau)d\tau. \tag{IF3}$$

The proof is nearly the same as for formulas (IF1) where $\psi = 0$ and (IF2) where $\psi = \pi/2$.

However, one has to check—depending on the choice of the basis function u_0 and the value of the impedance (or γ)—whether assumptions A_1 and A_2 are fulfilled and whether all singularities are located outside of the physical domain $z \geq 0$. In our opinion, the representations (IF1) and (IF2) cover all cases of physical importance and thus, we will not use (IF3) in the following.

In [Che10], Chen and Waubke also used complex integration and similar integration paths for solving the 2D problem (cp. our formula (IF3) and Fig. 2 in [Che10]).

Clearly, the convergence behavior of the line integral in (IF1) or (IF2) depends on $e^{\gamma\tau}$ or $e^{i\gamma\tau}$ and on $u_0(-z-\tau)$ or $u_0(-z-i\tau)$ for $\tau \to \infty$, respectively.

Let us, for example, consider the most important case of the sound field generated by a monopole above an impedance plane in 3D. Then we have to solve the boundary value problem for the Helmholtz equation (see [Och04])

$$W(u) := \Delta u + k^2 u = -\delta(x, y, z-h), \quad h > 0, z > 0 \tag{5.4a}$$

$$\frac{\partial u}{\partial z} + \gamma u = 0 \quad \text{for } z = 0 \text{ with } \gamma = \frac{ik}{Z_0}, \tag{5.4b}$$

where u has to satisfy the Sommerfeld radiation condition at infinity. Here, $\gamma = ik/Z_0$ with the normalized acoustical impedance $Z_0 = Z/(\rho c)$ at the impedance plane, and ρ is the fluid density.

Clearly, as basis solution the real monopole in free space can be chosen

$$u_0(z) = u_0(x, y, z; h) = \frac{1}{4\pi r(h)} e^{ikr(h)} \text{ with } r(h) = \sqrt{x^2 + y^2 + (z-h)^2}, \tag{5.5}$$

and formula (IF1) immediately gives the solution of the half-space-impedance problem

$$\begin{aligned}\hat{u}(z) &= u_0(z) + u_0(-z) + 2\gamma \int_0^\infty e^{\gamma\tau} u_0(-z-\tau)d\tau \\ &= \frac{1}{4\pi r(h)} e^{ikr(h)} + \frac{1}{4\pi r(-h)} e^{ikr(-h)} + 2\gamma \int_0^\infty e^{\gamma\tau} \frac{1}{4\pi r(-h-\tau)} e^{ikr(-h-\tau)} d\tau\end{aligned} \tag{5.6}$$

which is identical with Eq. (18) of [Och04]. The line integral

$$\sigma = 2\gamma \int_0^\infty e^{\gamma\tau} \frac{1}{4\pi r(-h-\tau)} e^{ikr(-h-\tau)} d\tau \tag{5.7}$$

can be considered as a correction term which is added to the sound field

$$u_{\text{rigid}} = \frac{1}{4\pi r(h)} e^{ikr(h)} + \frac{1}{4\pi r(-h)} e^{ikr(-h)} \tag{5.8}$$

generated by a monopole above a perfectly rigid plane. It is obvious that the integral (5.7) converges to a finite value only, if

$$\text{Re}\{\gamma\} < 0 \Leftrightarrow \text{Im}\{Z\} < 0, \tag{5.9a}$$

since

$$\gamma = \frac{ik}{Z_0} = \frac{ik(\text{Re}\{Z_0\} - i\text{Im}\{Z_0\})}{|Z_0|^2} = \frac{k\text{Im}\{Z_0\} + ik\text{Re}\{Z_0\}}{|Z_0|^2}, \tag{5.9b}$$

where Re{} or Im{} denotes the real or imaginary part of the quantity in brackets, respectively. Thus, by taking into account the time dependence $\exp(-i\omega t)$, the normal impedance of the impedance plane is enforced to be mass-like, e.g., $Z = \rho c Z_0 = -i\omega m$ for a mass m per unit area assigned to the impedance surface. Also, it can be seen that the condition (5.3a) for the basis function (5.5) reads

$$e^{\gamma\tau} \frac{e^{ikr(-\tau-h)}}{r(-\tau-h)} = e^{i\frac{k}{Z_0}\tau} \frac{e^{ik\sqrt{x^2+y^2+(z+h+\tau)^2}}}{\sqrt{x^2+y^2+(z+h+\tau)^2}} \underset{\tau\to\infty}{\longrightarrow} 0 \tag{5.10}$$

and is satisfied only for mass-like impedances (due to r in the denominator, real impedances are also allowed). On the other hand, it has been known for many years that surfaces with spring-like impedances can lead to the occurrence of surface waves. This topic was extensively discussed in the literature [Att02], [Emb96], [Mec89], [Ras89] starting with the work of Sommerfeld ([Som09], [Som26], [Som78]). The approach (5.6) seems not to be adequate to model such surface waves, which may be responsible for the divergent behavior of the line integral. However, the imaginary part of the ground impedance (the reactance) is positive for widely used models: For example, Embleton [Emb96] stated that "*for porous ground this imaginary component of the impedance is a compliance, like an elastic cushion*" (see [Emb96], p. 38). As a second example, the model of Delaney and Bazley (see [Att02], p. 74) predicts a positive reactance.

This gives the motivation to use the impedance formula (IF2) instead of (IF1), since the line integral over complex sources converges for all kind of surface impedances, which can be seen easily. By using (IF2), the solution

$$\hat{u}(z) = u_0(z) + u_0(-z) + 2i\gamma \int_0^\infty e^{i\gamma\tau} u_0(-z - i\tau)d\tau$$

$$= \frac{1}{4\pi r(h)} e^{ikr(h)} + \frac{1}{4\pi r(-h)} e^{ikr(-h)} + 2i\gamma \int_0^\infty e^{i\gamma\tau} \frac{1}{4\pi r(-h - i\tau)} e^{ikr(-h-i\tau)} d\tau \tag{5.11}$$

is obtained. Now, the line integral

$$\sigma = 2i\gamma \int_0^\infty e^{i\gamma\tau} \frac{1}{4\pi r(-h - i\tau)} e^{ikr(-h-i\tau)} d\tau$$
$$= 2i\gamma \int_0^\infty e^{-\frac{k}{Z_0}\tau} \frac{e^{ik\sqrt{x^2+y^2+(z+h+i\tau)^2}}}{\sqrt{x^2 + y^2 + (z + h + i\tau)^2}} d\tau \tag{5.12}$$

decays exponentially for real wave numbers k and $\tau \to \infty$, if

$$\mathrm{Re}\{i\gamma\} = \mathrm{Re}\left\{\frac{-k}{Z_0}\right\} < 0 \Leftrightarrow \mathrm{Re}\{Z_0\} > 0 \tag{5.13}$$

and then the condition (5.3b)

$$e^{i\gamma\tau} \frac{e^{ikr(-i\tau-h)}}{r(-i\tau - h)} = e^{-\frac{k}{Z_0}\tau} \frac{e^{ik\sqrt{x^2+y^2+(z+h+i\tau)^2}}}{\sqrt{x^2 + y^2 + (z + h + i\tau)^2}} \underset{\tau\to\infty}{\longrightarrow} 0 \tag{5.14}$$

is fulfilled.

The assumption (5.13) is true, if the surface impedance Z_0 possesses a small real part, which is a reasonable assumption, since acoustical propagation always involves dissipative processes. However, it is not a necessary condition, because the second factor of the expression (5.14) also decays exponentially, thereby allowing pure imaginary surface impedances, too. This fact can be verified by observing that the square root defining the radius in Eq. (5.14) must represent the root with a positive real part, in order to fulfill the radiation condition for the time dependency $\exp(-i\omega t)$. If we now choose receiver points at the z-axis ($x = y = 0$) for simplicity, we obtain for the exponential function in Eq. (5.14) ($z > 0, h > 0$)

$$e^{ik\sqrt{(z+h+i\tau)^2}} = e^{+ik(z+h)} e^{-k\tau} \underset{\tau\to+\infty}{\longrightarrow} 0. \tag{5.15}$$

Consequently, the improper integral in Eq. (5.12) exists for all kinds of surface impedances. Especially, the fact that we have to pass to a line integral consisting of complex monopoles instead of the normal real sources was the main motivation to study these complex monopoles in its own right. Clearly, at first glance, the transition from (IF1) to (IF2) can be performed by deforming the path of integration from the positive real axis to the imaginary axis in the complex τ-plane (see the Appendix in [Och13a]), and thus the complex sources seem to be only a mathematical construct without a deeper physical meaning. This was mentioned from Skudrzyk [Sku71] in Chap. XXV about the Sommerfeld Theory of Diffraction. In Sect. 25.8, p. 574, he investigated a spherical wave impinging on a straight edge, and wrote: "... R_α *must not be interpreted as a radius vector or distance but as a complex function of* α *with no physical meaning.*" However, the research work on complex sources in the acoustic and electromagnetic literature as described in Sect. 1.3 is very extensive and shows that nevertheless the complex source has a physical meaning that goes beyond the pure mathematical procedure of extending the path of integration from the real axis into the complex plane.

It is worth to note that Theorem 4 is not at all restricted to the Helmholtz equation. For example, Sommerfeld has found a solution of the heat equation (see [Som78]: pp. 59–61, Eq. (15)) with a mixed boundary condition which has exactly the same structure as (IF1). As basis solution, he has used the fundamental solution of the heat equation which is a Gaussian function ([Som78]: Eq. (7), p. (59)).

Half-space solutions of the Helmholtz equation in 3D will be considered in greater detail in Sect. 5.4 after having explored and tested the impedance formula for easier to handle cases in 1D and 2D, which we will study firstly.

5.2 One-Dimensional Green's Function on a Half-Line with Impedance Boundary Condition and Its Driving Sources

5.2.1 *Half-Line Green's Function*

As a first benchmark solution, we now consider the 1D Helmholtz equation

$$\frac{\mathrm{d}^2 p(z)}{\mathrm{d}z^2} + k^2 p(z) = -\delta(z-h) \quad \text{for } z > 0. \tag{5.16}$$

together with impedance boundary condition

$$\frac{p}{v_n} = Z \quad \text{at } z = 0. \tag{5.17}$$

Here, p is the complex sound pressure, v_n the normal velocity, and Z the normal impedance at $Z = 0$, where the unit normal is directed into the exterior of the acoustic domain V_+ along the negative z-axis. Since the gradient of the pressure is given by

$$\frac{\partial p}{\partial n} = i\omega\rho v_n \quad \text{at } z = 0, \tag{5.18}$$

the boundary condition (5.17) can be written as

$$\frac{\partial p}{\partial n} - \gamma p = 0 \text{ or } \frac{\partial p}{\partial z} + \gamma p = 0 \quad \text{at } z = 0. \tag{5.19}$$

Here again, $\gamma = ik/Z_0$ with the normalized acoustical impedance $Z_0 = Z/(\rho c)$. Now, we are looking for a solution of the 1D-Helmholtz equation (5.16) together with the boundary condition (5.19) in the "half-space" or better "half-line" V_+. Such a solution is also called a half-space Green's function, denoted now by $g_{1H} = p$. This 1D problem can be considered as a full 3D problem, in which only plane waves are propagating along the z-axis as shown in Fig. 5.1.

The solution of this problem is well-known and can be obtained by standard methods very easily, since we only deal with plane waves:

$$g_{1H}(z) = \frac{i}{2k}\left(\mathrm{e}^{ik|z-h|} + R\,\mathrm{e}^{ik|z+h|}\right). \tag{5.20}$$

where R is the reflection coefficient

$$R = \frac{Z_0 - 1}{Z_0 + 1}. \tag{5.21}$$

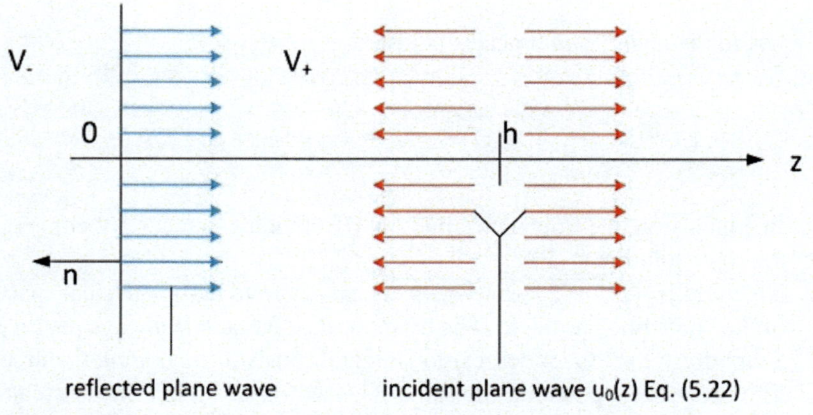

Fig. 5.1 The 1D plane wave "half-space problem"

(A) First, we try to get the same result by applying (IF1) from Theorem 4. Clearly, as basis function the Green's function in unlimited domain (cp. Eq. (4.4))

$$u_0(z) = g_1(z) = \frac{i}{2k} e^{ik|z-h|} \tag{5.22}$$

can be chosen. By introducing this function into formula (IF1), we obtain

$$\hat{u}(z) = \frac{i}{2k} e^{ik|z-h|} + \frac{i}{2k} e^{ik|z+h|} + \frac{i}{2k} \sigma_1^{(1)} \tag{5.23}$$

with

$$\sigma_1^{(1)} = 2\gamma \int\limits_0^\infty e^{\gamma\tau} e^{ik|z+h+\tau|} d\tau = 2\gamma e^{ikz_h} \int\limits_0^\infty e^{\gamma\tau} e^{ik\tau} d\tau \quad \text{for } z > 0 \tag{5.24}$$

with $z_h = z + h$.
This integral only converges under the assumption (5.9a) for a mass-like impedance with the result

$$\sigma_1^{(1)} = 2\gamma e^{ikz_h} e^{\tau(\gamma+ik)} \frac{1}{\gamma + ik}\bigg|_{\tau=0}^{\tau=\infty} = \frac{-2\gamma e^{ikz_h}}{\gamma + ik} \tag{5.25}$$

and therefore

$$\begin{aligned} \hat{u}(z) &= \frac{i}{2k} e^{ik|z-h|} + \frac{i}{2k} e^{ik|z+h|} - \frac{i}{2k} \frac{2\gamma e^{ikz_h}}{\gamma + ik} \\ &= \frac{i}{2k} e^{ik|z-h|} + \frac{i}{2k} e^{ik|z+h|} \left(1 - \frac{2\gamma}{\gamma + ik}\right) \end{aligned} \tag{5.26}$$

where the quantity in brackets is just

$$\left(1 - \frac{2\gamma}{\gamma + ik}\right) = \frac{-\gamma + ik}{\gamma + ik} = \frac{ik - \frac{ik}{Z_0}}{ik + \frac{ik}{Z_0}} = \frac{Z_0 - 1}{Z_0 + 1} = R. \tag{5.27}$$

In conclusion, the impedance integral (IF1) yields—in case of convergence—the right solution (5.20).
Note that $|z + h| = z + h = z_h$ is always positive in the physical space $z > 0$ of the "half-line" problem. The offset of the source h is always positive, too.

(B) Secondly, we use the superposition integral consisting of complex sources (IF2). The procedures go along the same lines as for (IF1), but now instead of $\sigma_1^{(1)}$ we have to use

$$\sigma_1^{(2)} = 2i\gamma \int_0^\infty e^{i\gamma\tau} e^{ik\sqrt{(z_h+i\tau)^2}} d\tau = 2i\gamma e^{ikz_h} \int_0^\infty e^{i\gamma\tau} e^{-k\tau} d\tau \quad \text{for } z > 0, \tag{5.28}$$

where the square root with positive real part has to be chosen, in order to fulfill the Sommerfeld radiation condition. Now, the integral converges for an arbitrary surface impedance Z (or γ) due to condition (5.13) with the same result as obtained from (IF1):

$$\sigma_1^{(2)} = 2i\gamma e^{ikz_h} e^{\tau(i\gamma-k)} \left. \frac{1}{i\gamma - k} \right|_{\tau=0}^{\tau=\infty} = \frac{-2i\gamma e^{ikz_h}}{i\gamma - k} = \sigma_1^{(1)}. \tag{5.29}$$

In summary, both formulas (IF1) and (IF2) give the correct solution in 1D for $z > 0$. However, for $z < 0$ the situation is different: (IF1) is a superposition with real plane waves in the lower half-space located at the half-line $z < -h$ as shown in Fig. 5.2.

In contrast, (IF2) is a superposition with "complex" plane waves

$$y(z_h; \tau) = \frac{i}{2k} e^{ik\sqrt{(z_h+i\tau)^2}} = \frac{i}{2k} e^{ik|z_h| - k\tau \, \mathrm{sgn}(z_h)}, \tag{5.30}$$

with jump singularities located at only one point $z = -h$. However, the magnitude of the jump increases with growing parameter τ. That is the reason that the integral—now considered for $-\infty < z_h < \infty$

$$\sigma_1^{(2)} = 2i\gamma \int_0^\infty e^{i\gamma\tau} e^{ik\sqrt{(z_h+i\tau)^2}} d\tau = 2i\gamma \int_0^\infty e^{i\gamma\tau} e^{ik|z_h| - k\tau \, \mathrm{sgn}(z_h)} d\tau \quad \text{for all } z \tag{5.31}$$

converges in the physical space $z_h > 0$, but not for every choice of γ in the image space $z_h < 0$.

This can be seen by looking upon the exponential function in the integral

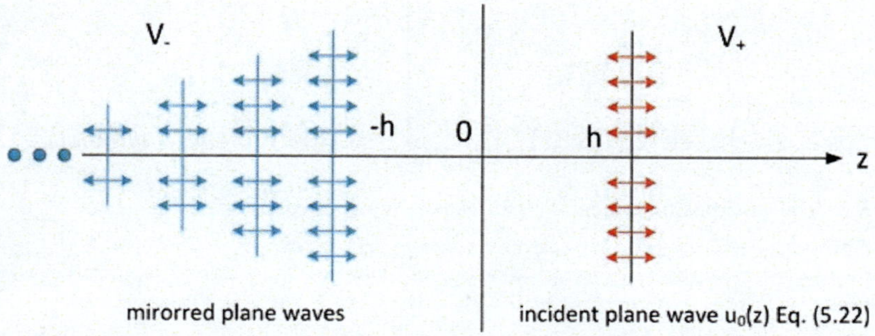

Fig. 5.2 The 1D plane wave "half-space problem"

$$E = \mathrm{e}^{i\gamma\tau}\mathrm{e}^{ik|z_h| - k\tau\,\mathrm{sgn}(z_h)} = \mathrm{e}^{i\gamma\tau}\mathrm{e}^{ik|z_h|}\begin{cases}\mathrm{e}^{-k\tau}, & z_h > 0\\ \mathrm{e}^{k\tau}, & z_h < 0\end{cases}$$

$$= \mathrm{e}^{ikZ_0''\tau/|Z_0|^2}\mathrm{e}^{ik|z_h|}\begin{cases}e^{-k\tau(Z_0'/|Z_0|^2+1)}, & z_h > 0\\ e^{-k\tau(Z_0'/|Z_0|^2-1)}, & z_h < 0\end{cases} \tag{5.32}$$

where we have taken into account that $i\gamma = -k/Z_0$ and have used the notation

$$Z_0' = \mathrm{Re}\{Z_0\},\ Z_0'' = \mathrm{Im}\{Z_0\},\ \text{ and } |Z_0| = \sqrt{Z_0'^2 + Z_0''^2}.$$

The surface impedance Z possesses a small real part, which is a reasonable assumption, since acoustical propagation always involves dissipative processes. Thus, in the acoustically relevant area $z_h > 0$, $E \underset{\tau\to\infty}{\longrightarrow} 0$, and the integral converges. However, in the mathematical image space $z_h < 0$, convergence demands that

$$\frac{Z_0'}{|Z_0|^2} - 1 > 0 \Leftrightarrow Z_0' > Z_0'^2 + Z_0''^2 \tag{5.33}$$

which seems to be very restrictive, and is satisfied, for example, if

$$0 < \mathrm{Re}\{Z_0\} < 1 \text{ and } \mathrm{Im}\{Z_0\} = 0. \tag{5.34}$$

Figure 5.3 shows symbolically the jumps of the plane waves Eq. (5.30) with increasing parameter τ.

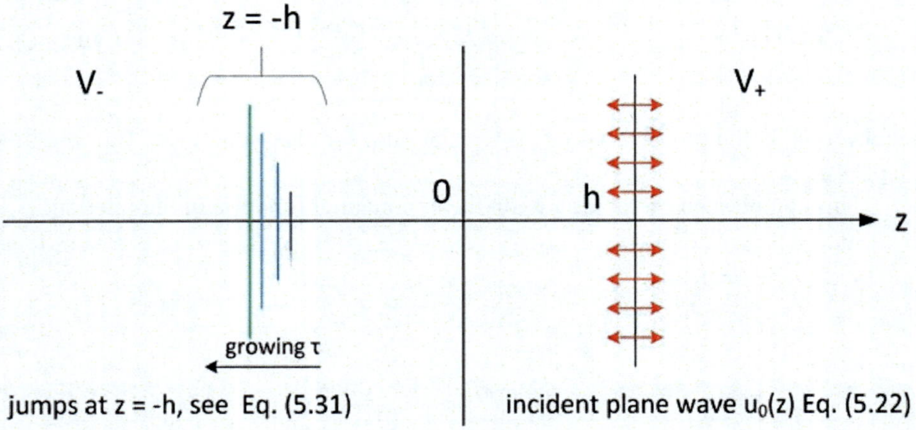

Fig. 5.3 The superposition integral (IF2) leads to jump singularities

5.2.2 *Driving Sources of the Half-Line Green's Function*

The physical meaning of these different representations of the "sigma correction" term is easier to understand when we consider the corresponding driving sources.

(A) We consider formula (IF1): By assuming that the Helmholtz-operator $D_1 := \frac{d^2}{dz^2} + k^2$ commutes with the line integral, we obtain from Eq. (5.23)

$$\begin{aligned} D_1\hat{u}(z) &= D_1\frac{i}{2k}e^{ik|z-h|} + D_1\frac{i}{2k}e^{ik|z+h|} + \frac{i}{2k}D_1\sigma_1^{(1)} \\ &= -\delta(z-h) - \delta(z+h) - 2\gamma\int_0^\infty e^{\gamma\tau}\delta(z_h+\tau)d\tau \end{aligned} \tag{5.35}$$

or

$$\begin{aligned} D_1\hat{u}(z) &= -\delta(z-h) - \delta(z+h) - 2\gamma\int_{-\infty}^{\infty} H(\tau)e^{\gamma\tau}\delta(z_h+\tau)d\tau \\ &= -\delta(z-h) - \delta(z+h) - 2\gamma H(-z_h)e^{-\gamma z_h} \end{aligned} \tag{5.36}$$

with the Heaviside function H.

Thus, the field of a plane wave above an impedance ground can be imagined as caused by a delta function plus an image delta function plus an exponential decreasing tail, if, for example, a mass-like impedance $Z = -i\omega m$ or $\gamma = -\rho/m$ is assumed (see Fig. 5.4).

(B) Now, we consider formula (IF2): Again we assume that the Helmholtz-operator $D_1 := \frac{d^2}{dz^2} + k^2$ commutes with the line integral, so that we get from Eq. (5.28) and Theorem 1 in Chap. 4

$$\begin{aligned} D_1\hat{u}(z) &= D_1\frac{i}{2k}e^{ik|z-h|} + D_1\frac{i}{2k}e^{ik|z+h|} + \frac{i}{2k}D_1\sigma_1^{(2)} \\ &= -\delta(z-h) - \delta(z+h) - 2i\gamma\int_0^\infty e^{i\gamma\tau}\tilde{\delta}_1(z_h,\tau)d\tau \end{aligned} \tag{5.37}$$

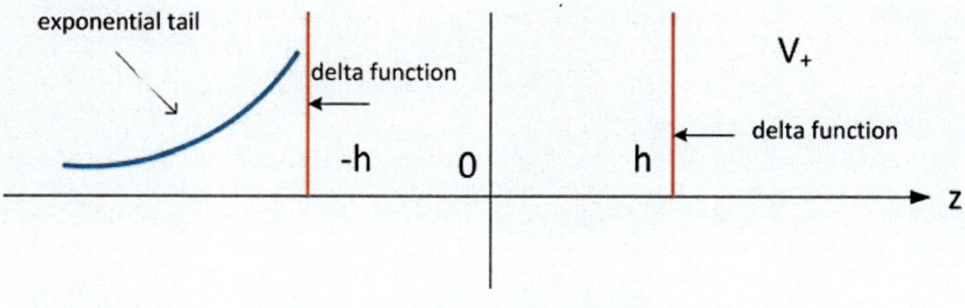

Fig. 5.4 Driving sources of the half-line Green's function using (IF1)

where the driving source of the complex monopole (4.11) comes into play:

$$\tilde{\delta}_1(z_h) = \tilde{\delta}_1(z_h, \tau) = \cosh(k\tau)\delta(z_h) + (i/k)\sinh(k\tau)\delta'(z_h). \tag{5.38}$$

Here, we have denoted the dependence of the complex driving source by τ explicitly. The integral appearing in Eq. (5.37) can be solved easily:

$$I_{1H} = 2i\gamma\big[A\delta(z_h) + B\delta'(z_h)\big] \tag{5.39}$$

with

$$A = \int\limits_0^\infty \cosh(k\tau)\mathrm{e}^{i\gamma\tau}\mathrm{d}\tau = \frac{i\gamma}{\gamma^2 + k^2} \tag{5.40}$$

and

$$B = \frac{i}{k}\int\limits_0^\infty \sinh(k\tau)\mathrm{e}^{i\gamma\tau}\mathrm{d}\tau = \frac{-i}{\gamma^2 + k^2} \tag{5.41}$$

under the necessary assumption for convergence that

$$\mathrm{Re}\{i\gamma \pm k\} < 0 \Rightarrow \mathrm{Re}\{Z_0\} > \mathrm{Re}^2\{Z_0\} + \mathrm{Im}^2\{Z_0\}, \tag{5.42}$$

which again is condition (5.33). That is not surprising, since the RHS of Eq. (5.37) is formulated for all $z \in (-\infty, \infty)$ and delivers the sources in the lower space, too.

By introducing A and B into Eq. (5.39), we obtain from Eq. (5.37) after some simple algebraic manipulations

$$\frac{\mathrm{d}^2\hat{u}}{\mathrm{d}z^2} + k^2\hat{u} = -\delta(z-h) - \frac{Z_0^2+1}{Z_0^2-1}\delta(z+h) - \frac{2iZ_0}{k\big(Z_0^2-1\big)}\delta'(z+h), \tag{5.43}$$

which is another representation for the driving source configuration of the impedance problem.

In summary, we have found three different representations of the driving sources located in the negative space $z \in (-\infty, 0)$:

(A) The ansatz with complex sources (IF2) gives

$$q_C = \delta(z-h) + \frac{Z_0^2+1}{Z_0^2-1}\delta(z+h) + \frac{2iZ_0}{k\left(Z_0^2-1\right)}\delta'(z+h), \quad Z_0 \neq 1. \tag{5.44}$$

(B) The ansatz with real sources (IF1) leads to (see Eq. (5.36))

$$q_R = \delta(z-h) + \delta(z+h) + 2\gamma H(-(z+h))\mathrm{e}^{-\gamma(z+h)}. \tag{5.45}$$

(C) The analytical solution (5.20) shows immediately that

$$q_A = \delta(z-h) + \frac{Z_0-1}{Z_0+1}\delta(z+h). \tag{5.46}$$

All three driving source configurations lead to the same sound field in the upper acoustically relevant space $z > 0$. For example, q_C leads to the solution (5.20), too, since (for $Z_0 \neq 1$)

$$\begin{aligned}\hat{u}(z) &= \frac{i}{2k}\left\{\mathrm{e}^{ik|z-h|} + \frac{Z_0^2+1}{Z_0^2-1}\mathrm{e}^{ik|z+h|} + \frac{2iZ_0}{k\left(Z_0^2-1\right)}\frac{\mathrm{d}}{\mathrm{d}z}\left(\mathrm{e}^{ik|z+h|}\right)\right\}\\ &= \frac{i}{2k}\left\{\mathrm{e}^{ik|z-h|} + \frac{Z_0-1}{Z_0+1}\mathrm{e}^{ik|z+h|}\right\}.\end{aligned} \tag{5.47}$$

For $z < 0$ one has to be careful, since the derivative of the absolute value $|z+h|$ leads to an additional Heaviside function! But clearly, in the mathematical space $z < 0$, the different source configurations generate different acoustical fields, which is of no physical relevance.

This is an illustrative example for the well-known fact that the inverse problem is not well posed. In other words, *the sources of a sound field are not uniquely determined by the sound field itself.*

Additionally, it is surprising that the result q_C (5.44) gives the correct driving sources for all impedances with the only limitation that $Z_0 \neq \pm 1$, although the derivation was only justified under the restrictive condition (5.42).

5.3 Two-Dimensional Green's Function on a Half-Plane with Impedance Boundary Condition and Its Driving Sources

5.3.1 Half-Plane Green's Function

We now want to solve the 2D Helmholtz equation

$$\frac{\partial^2 p(x,z)}{\partial x^2} + \frac{\partial^2 p(x,z)}{\partial z^2} + k^2 p(x,z) = -\delta(x, z-h) \tag{5.48}$$

together with impedance boundary condition

$$\frac{p}{v_{\rm n}} = Z \text{ or } \frac{\partial p}{\partial z} + \gamma p = 0 \quad \text{at } z = 0. \tag{5.49}$$

in the "half-space" or "half-plane" $V_+ = \{(x,z)|-\infty < x < \infty, z > 0\}$.

Here again, $\gamma = ik/Z_0$ with the normalized acoustical impedance $Z_0 = Z/(\rho c)$. The solution is also called a half-space Green's function in 2D, denoted now by $g_{2H} = p$. The resulting sound field can also be considered as generated by an infinite line source in the 3D half-space over an impedance plane as shown in Fig. 5.5.

The solution of this problem is not well-known and cannot be obtained as an integral-free expression analytically. Fortunately, the impedance formulas of Theorem 4 immediately give a solution.

Therefore, we apply formulas (IF1) and (IF2) from Theorem 4. Clearly, as basis function the 2D Green's function in unlimited domain (cp. Eq. (4.19))

$$u_0(x,z) = g_2(x,z) = \frac{i}{4} H_0^{(1)}\left(k\sqrt{x^2 + (z-z_0)^2}\right) \tag{5.50}$$

can be chosen. By introducing this function into formula (IF1) or (IF2), respectively, we obtain

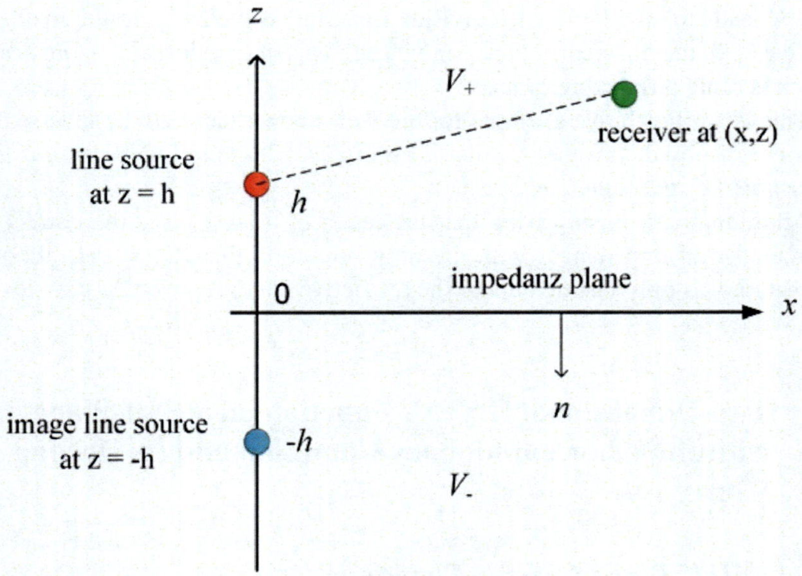

Fig. 5.5 Geometry of the 2D half-space problem

$$\hat{u}(z) = \frac{i}{4} H_0^{(1)}\left(k\sqrt{x^2 + (z-h)^2}\right) + \frac{i}{4} H_0^{(1)}\left(k\sqrt{x^2 + (z+h)^2}\right) + \sigma_2^{(1,2)} \quad (5.51)$$

with

$$\sigma_2^{(1)} = 2\gamma \int_0^\infty e^{\gamma\tau} \frac{i}{4} H_0^{(1)}\left(k\sqrt{x^2 + (z+h+\tau)^2}\right) d\tau \quad \text{for } z > 0 \quad (5.52)$$

or

$$\sigma_2^{(2)} = 2\gamma i \int_0^\infty e^{i\gamma\tau} \frac{i}{4} H_0^{(1)}\left(k\sqrt{x^2 + (z+h+i\tau)^2}\right) d\tau \quad \text{for } z > 0. \quad (5.53)$$

Note that Eq. (5.52) only converges for mass-like or purely absorbing impedances while Eq. (5.53) converges for all kind of impedances as explained at the end of Sect. 5.1. There are no closed-form solutions for these integrals, but we can compare it with other expressions found in the literature. As an example, we consider the solution (2.73) together with (2.75a, b) given by Attenborough et al. in [Att07] at pp. 57–58

$$\hat{u}(z) = \frac{i}{4} H_0^{(1)}\left(k\sqrt{x^2 + (z-h)^2}\right) + \frac{i}{4} H_0^{(1)}\left(k\sqrt{x^2 + (z+h)^2}\right) + I \quad (5.54)$$

with

$$I = \frac{-ik}{2\pi} \int_{-\infty}^{\infty} \frac{\beta}{k_z + k\beta} \frac{\exp(ik_z(z+h) + ik_x x)}{k_z} dk_x, \quad (5.55)$$

where $\gamma = ik\beta$ and $k_z = +\sqrt{k^2 - k_x^2}$. Making use of the identity

$$\frac{1}{k_z + k\beta} = \int_0^\infty e^{-\tau(k_z + k\beta)} d\tau \quad (5.56)$$

(convergence is assured due to the reasoning following Eq. (5.12) and the choice of the positive square root k_z) we obtain

$$I = \frac{-ik\beta}{2\pi} \frac{4\pi}{i} \int_0^\infty e^{-k\beta\tau} \left[\frac{i}{4\pi} \int_{-\infty}^{\infty} \frac{\exp(ik_z(z+h+i\tau) + ik_x x)}{k_x} dk_x \right] d\tau. \quad (5.57)$$

Due to the extended Sommerfeld integral in 2D Eq. (4.17), the term in square brackets can be expressed as a Hankel function

$$\frac{i}{4}H_0^{(1)}\left(k\sqrt{x^2+(z+h+i\tau)^2}\right)=\frac{i}{4\pi}\int\limits_{-\infty}^{\infty}\frac{\exp(ik_z(z+h+i\tau)+ik_x x)}{k_x}\mathrm{d}k_x \tag{5.58}$$

and the integral becomes

$$I=-2k\beta\int\limits_0^{\infty}\mathrm{e}^{-k\beta\tau}\frac{i}{4}H_0^{(1)}\left(k\sqrt{x^2+(z+h+i\tau)^2}\right)\mathrm{d}\tau=\sigma_2^{(2)}. \tag{5.59}$$

Since $-k\beta=i\gamma$, the solution of Attenborough et al. is the same as given in Eqs. (5.51) and (5.53) provided by the impedance formula (IF2).

Expressions (5.51) and (5.52) using the superposition with real sources (IF1) can be found in the thesis of Hein Hoernig (see [Hei10], p. 80, Eq. (3.8.2)).

5.3.2 *Driving Sources of the Half-Plane Green's Function*

We now determine the corresponding driving sources for the 2D half-plane problem.

(A) We consider formula (IF1) with real sources:
By assuming that the Helmholtz-operator $D_2:=\frac{\partial^2}{\partial x^2}+\frac{\partial^2}{\partial z^2}+k^2$ commutes with the line integral and using as basis function the 2D Green's function in unlimited domain $u_0(x,z)$ (cp. Eq. (4.19))

$$u_0(x,z;z_0)=g_2(x,z;h)=\frac{i}{4}H_0^{(1)}\left(k\sqrt{x^2+(z-z_0)^2}\right), \tag{5.60}$$

we obtain from Eqs. (5.51) and (5.52)

$$\begin{aligned}D_2\hat{u}(x,z)&=D_2u_0(x,z;h)+D_2u_0(x,z;-h)+2\gamma\int\limits_0^{\infty}\mathrm{e}^{\gamma\tau}D_2u_0(x,z;-h-\tau)\mathrm{d}\tau\\&=-\delta(x,z-h)-\delta(x,z+h)-\delta(x)2\gamma\int\limits_0^{\infty}\mathrm{e}^{\gamma\tau}\delta(z+h+\tau)\mathrm{d}\tau\end{aligned} \tag{5.61}$$

or

$$D_2\hat{u}(x,z) = -\left\{\delta(x, z-h) + \delta(x, z+h) + 2\gamma H(-z-h)\mathrm{e}^{-\gamma(z+h)}\right\}\delta(x) \tag{5.62a}$$

with the Heaviside function H. Clearly, this is nearly the same result as in 1D: The field in the upper space above an impedance ground can be imagined as caused by a delta function plus an image delta function plus an exponential decreasing tail. Using the 3D Helmholtz-operator and as basis function $u_0(x, y, z; z_0)$ the 3D Green's function (5.5) in unlimited domain, we again obtain nearly the same result. In conclusion, the driving sources in 1D, 2D, and 3D only differ by an additional multiplied Dirac delta function depending on the dimension of the problem ($z_h = z + h$):

$$\text{Driving sources in 1D: } S_1 = \delta(z-h) + \delta(z+h) + 2\gamma H(-z_h)\mathrm{e}^{-\gamma z_h}, \tag{5.62b}$$

$$\text{Driving sources in 2D: } S_2 = \left\{\delta(x, z-h) + \delta(x, z+h) + 2\gamma H(-z_h)\mathrm{e}^{-\gamma z_h}\right\}\delta(x), \tag{5.62c}$$

$$\text{Driving sources in 3D: } S_3 = \left\{\delta(x, y, z-h) + \delta(x, y, z+h) + 2\gamma H(-z_h)\mathrm{e}^{-\gamma z_h}\right\}\delta(x)\delta(y). \tag{5.62d}$$

(B) We now consider formula (IF2) with complex sources:
However, due to Theorem 2 in Chap. 4, the integral over the complex driving sources is more cumbersome and the results will strongly vary with the dimension of the problem.
Again we assume that the 2D Helmholtz-operator commutes with the line integral, so that we get from Eqs. (5.51) and (5.53)

$$\begin{aligned} D_2\hat{u}(x,z) &= D_2 g_2(x,z;h) + D_2 g_2(x,z;-h) + D_2\sigma_2^{(2)} \\ &= -\delta(x, z-h) - \delta(x, z+h) - 2i\gamma \int_0^\infty \mathrm{e}^{i\gamma\tau}\tilde{\delta}_2(x, z_h; \tau)\mathrm{d}\tau \end{aligned} \tag{5.63}$$

where due to Eq. (4.22) the driving source of the complex monopole in 2D comes into play:

$$\tilde{\delta}_2(x, z_h; \tau) = M_2(\tau)\delta(z_h) + iD_2(\tau)\delta'(z_h). \tag{5.64}$$

The dependence of the complex driving source on τ is denoted explicitly, since it is the imaginary part of the source location on the one hand and the integration variable on the other hand. The integral appearing in Eq. (5.63) can be solved under a suitable condition of convergence resulting in

$$\chi_2^{(2)} := -D_2\sigma_2^{(2)} = 2i\gamma \int_0^\infty \mathrm{e}^{i\gamma\tau}\tilde{\delta}_2(x, z_h; \tau)\mathrm{d}\tau = -\frac{\mathrm{e}^{ik\sqrt{1-\frac{1}{Z_0^2}}|x|}}{Z_0\sqrt{1-1/Z_0^2}}\left\{\frac{ik}{Z_0}\delta(z_h) - \delta'(z_h)\right\}. \tag{5.65}$$

The complete proof is given in Appendix B (see Eq. (B8)). Thus, the driving sources for the half-plane problem can also be represented as

$$S_2 = \delta(x, z-h) + \delta(x, z+h) - \frac{\mathrm{e}^{ik\sqrt{1-\frac{1}{Z_0^2}}|x|}}{Z_0\sqrt{1-1/Z_0^2}}\left\{\frac{ik}{Z_0}\delta(z_h) - \delta'(z_h)\right\}. \tag{5.66}$$

It is interesting to compare this result with the 1D driving sources (5.44) written in the form

$$S_1 = q_C = \delta(z-h) + \delta(z+h) - \frac{2\gamma}{\gamma^2 + k^2}\left\{\frac{ik}{Z_0}\delta(z_h) - \delta'(z_h)\right\}. \tag{5.67}$$

After having obtained the analogous result in 3D, a common formula for all three dimensions will be given.

5.4 Three-Dimensional Green's Function on a Half-Space with Impedance Boundary Condition and Its Driving Sources

5.4.1 *Half-Space Green's Function*

We now want to solve the 3D Helmholtz equation

$$\begin{aligned} D_3 p(x, y, z) &:= \frac{\partial^2 p(x, y, z)}{\partial x^2} + \frac{\partial^2 p(x, y, z)}{\partial y^2} + \frac{\partial^2 p(x, y, z)}{\partial z^2} + k^2 p(x, y, z) \\ &= -\delta(x, y, z-h) \end{aligned} \tag{5.68}$$

together with impedance boundary condition

$$\frac{p}{v_\mathrm{n}} = Z \text{ or } \frac{\partial p}{\partial z} + \gamma p = 0 \quad \text{at } z = 0, \tag{5.69}$$

where $V_+ = \{(x, y, z)| -\infty < x < \infty, -\infty < y < \infty, z > 0\}$ is now a real half-space.

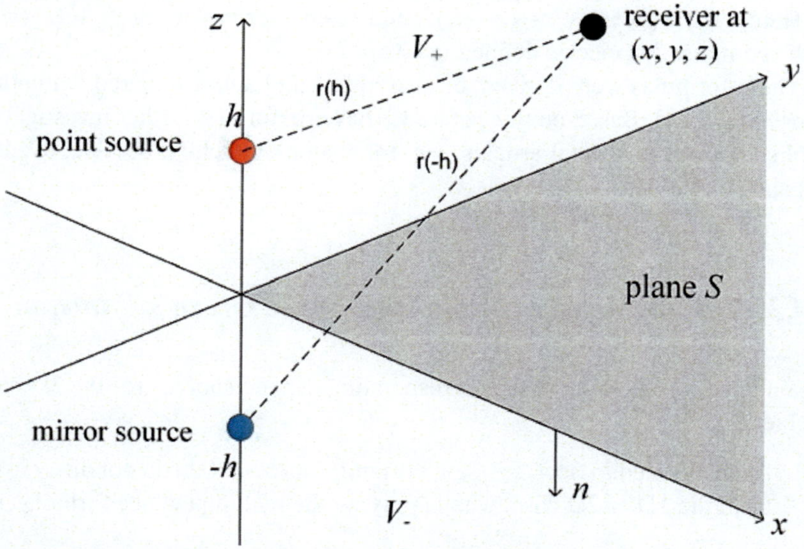

Fig. 5.6 Geometry of the 3D half-space problem

Here again, $\gamma = ik/Z_0$ with the normalized acoustical impedance $Z_0 = Z/(\rho c)$. The solution is also called a half-space Green's function in 3D, denoted now by $g_{3H} = p$. The geometry of the 3D half-space problem is shown in Fig. 5.6.

As well as in 2D, the solution cannot be obtained as an integral-free closed-form expression, but the impedance formulas of Theorem 4 immediately give a solution.

Therefore, we apply formulas (IF1) and (IF2) from Theorem 4 as it was performed directly after the proof of Theorem 4 to illustrate the impedance formulas by a practical example in 3D space. By using (IF1) or (IF2), we obtain (cp. Eqs. (5.6) and (5.11))

$$\hat{u}(z) = g_{H3}(x, y, z) = \frac{1}{4\pi r(h)} \mathrm{e}^{ikr(h)} + \frac{1}{4\pi r(-h)} \mathrm{e}^{ikr(-h)} + \sigma_3^{(1,2)} \tag{5.70}$$

with

$$\sigma_3^{(1)} = 2\gamma \int_0^\infty \mathrm{e}^{\gamma\tau} \frac{1}{4\pi r(-h-\tau)} \mathrm{e}^{ikr(-h-\tau)} \mathrm{d}\tau \tag{5.71}$$

or

$$\sigma_3^{(2)} = 2i\gamma \int_0^\infty \mathrm{e}^{i\gamma\tau} \frac{1}{4\pi r(-h-i\tau)} \mathrm{e}^{ikr(-h-i\tau)} \mathrm{d}\tau. \tag{5.72}$$

Here, $r(h)=\sqrt{x^2+y^2+(z-h)^2}$, and $r(-h-i\tau)=\sqrt{x^2+y^2+(z+h+i\tau)^2}$ is the complex distance as defined in Chap. 2.

These formulas were derived, discussed in detail, and compared with other solutions in [Och04]. Based on this paper, we have performed further investigations and published them in several articles, see, for example, [Och05, 06, 08c, 09, 13a, 13b, 16] and [Pis17, 19].

5.4.2 Driving Sources of the Half-Space Green's Function

Again, we try to determine the corresponding driving sources for the 3D half-plane problem.

(A) As mentioned in Sect. 5.3.2, the driving sources in 3D do not differ essentially from the 1D or 2D case, when applying the real impedance formula (IF1):

$$S_3=\left\{\delta(x,y,z-h)+\delta(x,y,z+h)+2\gamma H(-z_h)\mathrm{e}^{-\gamma z_h}\right\}\delta(x)\delta(y). \tag{5.73}$$

(B) However, due to Theorem 2 in Chap. 4, the integral over the complex driving sources is more difficult to solve, when using formula (IF2). Again we assume that the 3D Helmholtz-operator commutes with the line integral, so that we get from Eqs. (5.70) and (5.72)

$$\begin{aligned} D_3\hat{u}(x,y,z) &= D_3g_3(x,y,z;h)+D_3g_3(x,y,z;-h)+D_3\sigma_3^{(2)} \\ &= -\delta(x,y,z-h)-\delta(x,y,z+h)-2i\gamma\int\limits_0^\infty \mathrm{e}^{i\gamma\tau}\tilde{\delta}_3(x,y,z_h;\tau)\mathrm{d}\tau \end{aligned} \tag{5.74}$$

where due to Eq. (4.51) the driving source of the complex monopole in 3D comes into play:

$$\tilde{\delta}_3(x,y,z_h;\tau)=M_3(\tau)\delta(z_h)+iD_3(\tau)\delta'(z_h). \tag{5.75}$$

Again, the dependence of the complex driving source on τ is denoted explicitly. The integral appearing in Eq. (5.74) can be solved under a suitable condition of convergence resulting in

$$\begin{aligned} \chi_3^{(2)} &:= -D_3\sigma_3^{(2)} = 2i\gamma\int\limits_0^\infty \mathrm{e}^{i\gamma\tau}\tilde{\delta}_3(x,z_h;\tau)\mathrm{d}\tau \\ &= -\frac{k}{2Z_0}H_0^{(1)}\left(kr\sqrt{1-\frac{1}{Z_0^2}}\right)\left\{\frac{ik}{Z_0}\delta(z_h)-\delta'(z_h)\right\}. \end{aligned} \tag{5.76}$$

The complete proof is given in Appendix C. Thus, the driving sources for the half-space problem can also be represented as

$$S_3 = \delta(x, y, z-h) + \delta(x, y, z+h) - \frac{k}{2Z_0} H_0^{(1)}\left(kr\sqrt{1-\frac{1}{Z_0^2}}\right)\left\{\frac{ik}{Z_0}\delta(z_h) - \delta'(z_h)\right\} \tag{5.77}$$

with $r = \sqrt{x^2+y^2}$.

Now, we can compare the driving sources for all three dimensions.

5.5 Overview About the Sources and a Generalized Formula for 1D, 2D, and 3D

For getting a good and easy overview, we first put together the formulas for Green's functions in different dimensions:

(a) In 3D the normal free-field Green's function

$$g_3(x, y, z; k) = \frac{1}{4\pi\sqrt{x^2+y^2+z^2}} \mathrm{e}^{ik\sqrt{x^2+y^2+z^2}} \tag{5.78}$$

is the solution of the 3D Helmholtz equation

$$\begin{aligned} \Delta_3 g_3 + k^2 g_3 &= \frac{\partial^2 g_3(x, y, z)}{\partial x^2} + \frac{\partial^2 g_3(x, y, z)}{\partial y^2} + \frac{\partial^2 g_3(x, y, z)}{\partial z^2} + k^2 g_3 \\ &= -\delta(x, y, z). \end{aligned} \tag{5.79}$$

(b) In 2D the free-field Green's function

$$g_2(x, z; k) = \frac{i}{4} H_0^{(1)}\left(k\sqrt{x^2+z^2}\right) \tag{5.80}$$

is the solution of the 2D Helmholtz equation

$$\Delta_2 g_2 + k^2 g_2 = \frac{\partial^2 g_2}{\partial x^2} + \frac{\partial^2 g_2}{\partial z^2} + k^2 g_2 = -\delta(x, z). \tag{5.81}$$

(c) In 1D the Green's function

$$g_1(z; k) = \frac{i}{2k} \mathrm{e}^{ik|z|} \tag{5.82}$$

is the solution of the 1D Helmholtz equation

$$\Delta_1 g_1 + k^2 g_1 = \frac{\mathrm{d}^2 g_1}{\mathrm{d}z^2} + k^2 g_1 = -\delta(z). \tag{5.83}$$

Please do not confuse these Green's functions with the complex monopoles in the previous chapters which also are denoted by g_1, g_2, g_3. What is meant is easy to understand from the context.

(d) However, what happens for the "Helmholtz equation of dimension zero"? Clearly, then all derivatives and independent variables must disappear and it remains (by our definition)

$$\Delta_0 g_0 + k^2 g_0 = k^2 g_0 = -1. \tag{5.84}$$

Thus we get the simple solution for the "0D Green's function"

$$g_0(k) = -\frac{1}{k^2}. \tag{5.85}$$

It is interesting to see, how the singularity of the Green's function grows from dimension to dimension.

Now we consider the driving sources for the complex Green's functions, starting with the formulation in 1D. From Eqs. (5.67) and (5.85) we obtain ($\gamma = ik/Z_0$)

$$S_1 = \delta(z-h) + \delta(z+h) + 2\gamma g_0(\tilde{\kappa})\{\gamma\delta(z_h) - \delta'(z_h)\} \tag{5.86}$$

with $\tilde{\kappa}^2 = k^2 + \gamma^2 = k^2(1 - 1/Z_0^2)$.

In 2D, we obtain from Eqs. (5.66) and (5.82)

$$S_2 = \delta(x, z-h) + \delta(x, z+h) + 2\gamma g_1(x; \tilde{\kappa})\{\gamma\delta(z_h) - \delta'(z_h)\}. \tag{5.87}$$

And in 3D Eqs. (5.77) and (5.80) lead to

$$S_3 = \delta(x, y, z-h) + \delta(x, y, z+h) + 2\gamma g_2(x, y; \tilde{\kappa})\{\gamma\delta(z_h) - \delta'(z_h)\}. \tag{5.88}$$

In summary, we obtain the beautiful formula

$$S_m = \delta_m(z-h) + \delta_m(z+h) + 2\gamma g_{m-1}(\tilde{\kappa})\{\gamma\delta(z_h) - \delta'(z_h)\} \tag{5.89}$$

for $m = 1, 2, 3$, where $\delta_m(z)$ denotes the Dirac delta function of dimension m (e.g., $\delta_3(z) = \delta(x, y, z)$).

We suspect that Eq. (5.89) is also valid in four-dimensional space, since we can use it together with Eq. (5.78), but that obviously has no physical meaning.

Also, we have to mention that these formulas have been derived by assuming the validity of condition (5.42) in order to guarantee the necessary convergence behavior of the integrals. For example, convergence is ensured, if condition (5.34)

$$0 < \mathrm{Re}\{Z_0\} < 1 \text{ and } \mathrm{Im}\{Z_0\} = 0$$

is fulfilled. But then, the wave number $\tilde{\kappa} = k\sqrt{(1 - 1/Z_0^2)}$ becomes imaginary, and $g_1(z; \kappa)$ and $g_2(x, z; \kappa)$ must be evaluated for a complex wave number. It is not clear, whether Eq. (5.89) is still valid in such a case. These mathematical aspects of convergence and appearance of "complex wave numbers" need further investigation. Nevertheless, the formulas for the driving sources make it possible to derive solutions of the half-space impedance problem that agree with well-known solutions from the literature.

Chapter 6
New and Old Formulas from the Helmholtz Equation with Half-Space Driving Sources

Now, the question arises, what benefit does the knowledge of the driving sources have. One application is that they allow to derive new solutions of the half-space impedance problem by solving the corresponding inhomogeneous Helmholtz equation in free space instead of the original half-space problem with additional boundary condition at a certain infinite plane, for example, at $z = 0$, which was our choice in the preceding chapters.

For comparison and validation with the new formulas that we will derive in this chapter, the solutions can be used that have already been obtained with the superposition integral given in Theorem 4 in Chap. 5 (see Eqs. (5.70)–(5.72) for the 3D case).

6.1 Solutions Obtained by Using the Helmholtz Huygens Integral

It is well-known that the solution of the inhomogeneous Helmholtz equation (see, e.g., [Mor53], part I, p. 806, Eq. 7.28 or [Sku71], p. 495, Eq. (28)) can be obtained by a convolution integral of the RHS and a suitable Green's function. For example, the 3D Helmholtz equation together with the "real" source contribution (5.73)

$$\Delta_3 G_3 + k^2 G_3 = -S_3 \tag{6.1}$$

is solved by

$$G_3(x, y, z) = \iiint\limits_{R^3} S_3(\vec{y}) \frac{e^{ik|\vec{x}-\vec{y}|}}{4\pi|\vec{x}-\vec{y}|} d\vec{y}$$

M. Ochmann and R. Piscoya, *Theory and Application of Acoustic Sources Using Complex Analysis*, https://doi.org/10.1007/978-981-33-6040-2_6

$$= \frac{e^{ik\sqrt{x^2+y^2+(z-h)^2}}}{4\pi\sqrt{x^2+y^2+(z-h)^2}} + \frac{e^{ik\sqrt{x^2+y^2+(z+h)^2}}}{4\pi\sqrt{x^2+y^2+(z+h)^2}}$$

$$+ \frac{2ik}{Z_0}\int_{-\infty}^{-h} e^{-\frac{ik}{Z_0}(z_0+h)} \frac{e^{ik\sqrt{x^2+y^2+(z-z_0)^2}}}{4\pi\sqrt{x^2+y^2+(z-z_0)^2}} dz_0 \qquad (6.2)$$

where as usual $\vec{x} = (x, y, z)^{\mathrm{T}}$ is the field point and $\vec{y} = (x_0, y_0, z_0)^{\mathrm{T}}$ is the source location, T denotes transposition. In the sequel, the half-space Green's functions will be denoted by capital letters, in order to distinguish them from the free-field Green's functions (with and without complex source location). By using the substitution $z_0 = -\tau - h$, it becomes evident that Eq. (6.2) is identical with solution (5.6) which was obtained by applying the real superposition integral from Theorem 4 in Chap. 5. As mentioned below Eq. (5.6), this solution only converges for real or mass-like impedances, but for instance, not for a spring-like impedance. Using instead of the "real" source contribution the "complex" driving sources S_3 (5.77) and inserting them into Eq. (6.2), we get immediately due to the properties of the Dirac delta function

$$G_3(x, y, z) = \frac{1}{4\pi}\frac{e^{ik\sqrt{x^2+y^2+(z-h)^2}}}{\sqrt{x^2+y^2+(z-h)^2}} + \frac{1}{4\pi}\frac{e^{ik\sqrt{x^2+y^2+(z+h)^2}}}{\sqrt{x^2+y^2+(z+h)^2}}$$

$$- \frac{k}{2Z_0}\iint_{R^2} H_0^{(1)}\left(kr_0\sqrt{1-\frac{1}{Z_0^2}}\right)\cdot\left\{\frac{ik}{Z_0}\frac{e^{ik\sqrt{(x-x_0)^2+(y-y_0)^2+(z+h)^2}}}{4\pi\sqrt{(x-x_0)^2+(y-y_0)^2+(z+h)^2}}\right.$$

$$\left.-\frac{\partial}{\partial z}\frac{e^{ik\sqrt{(x-x_0)^2+(y-y_0)^2+(z+h)^2}}}{4\pi\sqrt{(x-x_0)^2+(y-y_0)^2+(z+h)^2}}\right\} dx_0 dy_0 \qquad (6.3a)$$

with $r_0 = \sqrt{x_0^2 + y_0^2}$.

This result is exactly the layer potential representation, which Habault and Filippi gave in their paper [Hab81], Eqs. (17)–(20) or in the text book [Fil99] at p. 124, Eq. (4.6). Note that the different minus sign stems from the different definition of the RHS of the Helmholtz equation. Thus, the layer potential representation and formula (5.11) based on superposition integral with complex sources are equivalent.

By applying the same procedure to the 2D Helmholtz equation

$$\Delta_2 G_2 + k^2 G_2 = -S_2, \qquad (6.3b)$$

we get the convolution integral

$$G_2(x, z) = \iint_{R^2} S_2(\vec{y})\frac{i}{4}H_0^{(1)}(k|\vec{x}-\vec{y}|)d\vec{y}, \qquad (6.3c)$$

where now $\vec{x} = (x, z)^{\mathrm{T}}$ is the field point and $\vec{y} = (x_0, z_0)^{\mathrm{T}}$ is the source location.

By using the real 2D-sources S_2 of Eq. (5.62c), we get formula (5.51) together with (5.52) which suffers from the convergence problems described above.

By introducing the complex 2D-sources (5.66) into Eq. (6.3c), we get the result (cp. Eq. (6.3a))

$$G_2(x,z) = \frac{i}{4} H_0^{(1)}\left(k\sqrt{x^2+(z-h)^2}\right) + \frac{i}{4} H_0^{(1)}\left(k\sqrt{x^2+(z+h)^2}\right)$$
$$-\int_{-\infty}^{\infty} \frac{\mathrm{e}^{ik\sqrt{1-\frac{1}{Z_0^2}}|x_0|}}{Z_0\sqrt{1-1/Z_0^2}} \left\{ \frac{ik}{Z_0}\frac{i}{4} H_0^{(1)}\left(k\sqrt{(x-x_0)^2+(z+h)^2}\right)\right.$$
$$\left. -\frac{\partial}{\partial z}\frac{i}{4} H_0^{(1)}\left(k\sqrt{(x-x_0)^2+(z+h)^2}\right)\right\} \mathrm{d}x_0 \tag{6.4}$$

This formula seems to be new, but unfortunately it has a relatively complicated structure.

6.2 Solutions by Fourier Transform in *z* Followed by Fourier or Hankel Transform

In this chapter, we perform the derivation in 2D and in 3D separately, in order to keep an easier overview about the formulas.

6.2.1 2D Case: Solution by Fourier Transform in z *Followed by Fourier Transform in* x

1. The starting point is the 2D Helmholtz equation

$$\frac{\partial^2 G_2(x,z)}{\partial x^2} + \frac{\partial^2 G_2(x,z)}{\partial z^2} + k^2 G_2(x,z) = -\delta(x, z-h) - \delta(x, z-h) - \chi_2^{(2)} \tag{6.5}$$

with (see Eq. (5.65))

$$\chi_2^{(2)} = 2i\gamma \int_0^{\infty} \mathrm{e}^{i\gamma\tau} \tilde{\delta}_2(x, z_h; \tau)\mathrm{d}\tau = -\frac{\mathrm{e}^{ik\sqrt{1-\frac{1}{Z_0^2}}|x|}}{Z_0\sqrt{1-1/Z_0^2}} \left\{ \frac{ik}{Z_0}\delta(z_h) - \delta'(z_h) \right\}. \tag{6.6}$$

Taking into account the third term only, we get

$$\frac{\partial^2 h_2(x,z)}{\partial x^2} + \frac{\partial^2 h_2(x,z)}{\partial z^2} + k^2 h_2(x,z) = -\chi_2^{(2)} \tag{6.7}$$

where h_2 is the correction term of the solution for the sound rigid case.
As a first step, a Fourier transform F_z in z of Eq. (6.7) is carried out

$$\begin{aligned}\frac{\mathrm{d}^2\hat{h}_2}{\mathrm{d}x^2} + \left(k^2 - k_z^2\right)\hat{h}_2 &= \frac{\mathrm{e}^{ik\mu|x|}}{\mu Z_0}\left(\frac{ik}{Z_0}\int_{-\infty}^{\infty}\delta(z_h)\mathrm{e}^{-ik_z z}\mathrm{d}z - \int_{-\infty}^{\infty}\delta'(z_h)\mathrm{e}^{-ik_z z}\mathrm{d}z\right)\\ &= \frac{\mathrm{e}^{ik\mu|x|}}{\mu Z_0}\mathrm{e}^{ik_z h}\left(\frac{ik}{Z_0} - ik_z\right),\end{aligned} \tag{6.8}$$

where by definition

$$F_z[f(z)] = \hat{f}(k_z) = \int_{-\infty}^{\infty} f(z)\mathrm{e}^{-ik_z z}\mathrm{d}z \text{ and } \mu = \sqrt{1 - 1/Z_0^2}. \tag{6.9}$$

2. The second step is a Fourier transform in x of (6.8) defined by

$$F_x[g(x)] = \tilde{f}(k_x) = \int_{-\infty}^{\infty} g(x)\mathrm{e}^{-ik_x x}\mathrm{d}x, \tag{6.10}$$

which gives

$$\left(k^2 - k_x^2 - k_z^2\right)\tilde{\hat{h}}_2 = \frac{i\mathrm{e}^{ik_z h}}{\mu Z_0}\left(\frac{k}{Z_0} - k_z\right)\int_{-\infty}^{\infty}\mathrm{e}^{ik\mu|x|}\mathrm{e}^{-ik_x x}\mathrm{d}x \tag{6.11}$$

or

$$\tilde{\hat{h}}_2 = \frac{i\mathrm{e}^{ik_z h}}{\mu Z_0\left(k^2 - k_x^2 - k_z^2\right)}\left(\frac{k}{Z_0} - k_z\right)\int_{-\infty}^{\infty}\mathrm{e}^{ik\mu|x|}\mathrm{e}^{-ik_x x}\mathrm{d}x. \tag{6.12}$$

By choosing the branch of the square root such that $\operatorname{Im}\mu = \operatorname{Im}\sqrt{1 - 1/Z_0^2} > 0$, it is easy to see that the Fourier transform exists, and we find that

$$I := \int_{-\infty}^{\infty}\mathrm{e}^{ik\mu|x|}\mathrm{e}^{-ik_x x}\mathrm{d}x = \frac{2k\mu i}{(k\mu)^2 - k_x^2}. \tag{6.13}$$

This can be easily seen by considering the auxiliary problem

$$y'' + (k\mu)^2 y = -\delta(x) \text{ or } F_x[y] = \tilde{y} = \frac{-1}{(k\mu)^2 - k_x^2} \quad (6.14)$$

with known solution

$$y(x) = \frac{i}{2k\mu} e^{ik\mu|x|}. \quad (6.15)$$

It follows that

$$F_x\left[\frac{i}{2k\mu} e^{ik\mu|x|}\right] = \frac{-1}{(k\mu)^2 - k_x^2} = \frac{i}{2k\mu} I, \quad (6.16)$$

which confirms expression (6.13). Therefore,

$$\tilde{\hat{h}}_2 = \frac{2k\mu i}{\mu Z_0} i e^{ik_z h} (k/Z_0 - k_z) \frac{1}{k^2 - k_x^2 - k_z^2} \frac{1}{(k\mu)^2 - k_x^2} = w \frac{1}{k_x^2 - \alpha^2} \frac{1}{k_x^2 - \beta^2}, \quad (6.17)$$

where we have used the abbreviations ($\gamma = ik/Z_0$)

$$\alpha^2 = k^2 - k_z^2, \quad \beta^2 = (k\mu)^2, \quad w = 2\gamma i e^{ik_z h} (k/Z_0 - k_z). \quad (6.18)$$

By partial fraction decomposition, we obtain

$$\tilde{\hat{h}}_2 = w \left(\frac{1}{k_x^2 - \alpha^2} - \frac{1}{k_x^2 - \beta^2} \right) \frac{1}{\alpha^2 - \beta^2}. \quad (6.19)$$

3. The third step is the inverse Fourier transform in x. Again, by using the identity (6.16), we find that

$$\hat{h}_2 = F_x^{-1}\left[\tilde{\hat{h}}_2\right] = w \left(\frac{i}{2\alpha} e^{i\alpha|x|} - \frac{i}{2\beta} e^{i\beta|x|} \right) \frac{1}{\alpha^2 - \beta^2} \quad (6.20)$$

or equivalently by substituting back the original quantities defined in (6.18)

$$\hat{h}_2 = 2\gamma i e^{ik_z h} (k/Z_0 - k_z) \left(\frac{i}{2k\mu} e^{ik\mu|x|} - \frac{i}{2\sqrt{k^2 - k_z^2}} e^{i\sqrt{k^2 - k_z^2}|x|} \right) \frac{-1}{(k/Z_0)^2 - k_z^2}. \quad (6.21)$$

4. The final step is the inverse Fourier transform in z, which yields

$$h_2 = -\frac{i\gamma}{\pi} \int_{-\infty}^{\infty} \frac{1}{k/Z_0 + k_z} \left(\frac{i}{2k\sqrt{1 - 1/Z_0^2}} e^{ik\mu|x|} - \frac{i}{2k\sqrt{1 - k_z^2/k^2}} e^{i\sqrt{k^2 - k_z^2}|x|} \right) e^{ik_z(z+h)} dk_z \quad (6.22)$$

and finally

$$G_2(x,z) = \frac{i}{4}H_0^{(1)}\left(k\sqrt{x^2+(z-h)^2}\right) + \frac{i}{4}H_0^{(1)}\left(k\sqrt{x^2+(z+h)^2}\right)$$
$$-\frac{i\gamma}{\pi}\int_{-\infty}^{\infty}\frac{1}{k/Z_0+k_z}\left(\frac{i}{2k\sqrt{1-1/Z_0^2}}e^{ik\mu|x|} - \frac{i}{2k\sqrt{1-k_z^2/k^2}}e^{i\sqrt{k^2-k_z^2}|x|}\right)e^{ik_z(z+h)}dk_z. \tag{6.23}$$

This new formula has a simple structure. For instance, for a real impedance with $0 < Z_0 < 1$ we obtain that $\mu = \sqrt{1-1/Z_0^2} = i\mathrm{Im}\,\mu = i\mu''$ is pure imaginary with $\mu'' > 0$ due to the choice of the square root. Thus, for $|x| \to \infty$, the first term of h_2 disappears which gives

$$G_2(x,z) = \frac{i}{4}H_0^{(1)}\left(k\sqrt{x^2+(z-h)^2}\right) + \frac{i}{4}H_0^{(1)}\left(k\sqrt{x^2+(z+h)^2}\right)$$
$$+\frac{i\gamma}{\pi}\int_{-\infty}^{\infty}\frac{1}{k/Z_0+k_z}\frac{i}{2k\sqrt{1-k_z^2/k^2}}e^{i\sqrt{k^2-k_z^2}|x|}e^{ik_z(z+h)}dk_z.$$

The first term of h_2 also disappears for a perfectly soft condition, i.e., $Z_0 \to 0$ so that $\mu'' \to +\infty$. Thus, h_2 becomes

$$h_2 = \frac{-k}{\pi}\int_{-\infty}^{\infty}\frac{1}{k+Z_0k_z}\left(\frac{i}{2\sqrt{k^2-k_z^2}}e^{i\sqrt{k^2-k_z^2}|x|}\right)e^{ik_z(z+h)}dk_z.$$

Taking again into account that $Z_0 \to 0$ and the formula for the plane wave decomposition (4.16), one obtains

$$h_2 = \frac{-1}{\pi}\int_{-\infty}^{\infty}\frac{i}{2\sqrt{k^2-k_z^2}}e^{i\sqrt{k^2-k_z^2}|x|}e^{ik_z(z+h)}dk_z = -2\frac{i}{4}H_0^{(1)}\left(k\sqrt{x^2+(z+h)^2}\right)$$

and thus

$$G_2(x,z) = \frac{i}{4}H_0^{(1)}\left(k\sqrt{x^2+(z-h)^2}\right) - \frac{i}{4}H_0^{(1)}\left(k\sqrt{x^2+(z+h)^2}\right),$$

which is the correct solution for the reflection at a perfectly soft plane.

6.2.2 3D Case: Solution by Fourier Transform in z Followed by Hankel Transform in (x, y)

The solution procedure is nearly the same as in 2D. The only difference is that we now have to take a Hankel transform with respect to the variable $r = \sqrt{x^2 + y^2}$ as a second step and an inverse Hankel transform as third step.

1. The starting point is the inhomogeneous 3D Helmholtz equation

$$\frac{\partial^2 G_3}{\partial x^2} + \frac{\partial^2 G_3}{\partial y^2} + \frac{\partial^2 G_3}{\partial z^2} + k^2 G_3 = -\delta(x, y, z - h) - \delta(x, y, z + h) - \chi_3^{(2)} \tag{6.24}$$

with (see Eq. (5.76))

$$\chi_3^{(2)} = 2i\gamma \int_0^\infty \mathrm{e}^{i\gamma\tau} \tilde{\delta}_3(x, z_h; \tau)\mathrm{d}\tau = -\frac{k}{2Z_0} H_0^{(1)}\left(kr\sqrt{1 - \frac{1}{Z_0^2}}\right)\{\gamma\delta(z_h) - \delta'(z_h)\}. \tag{6.25}$$

Again, we only consider the third term

$$\frac{\partial^2 h_3(x, y, z)}{\partial x^2} + \frac{\partial^2 h_3(x, y, z)}{\partial y^2} + \frac{\partial^2 h_3(x, y, z)}{\partial z^2} + k^2 h_3(x, y, z) = -\chi_3^{(2)}, \tag{6.26}$$

which is the correction term of the solution for the sound rigid case.
As a first step, a Fourier transform F_z in z of Eq. (6.26) is carried out

$$\begin{aligned}\frac{\partial^2 \hat{h}_3}{\partial x^2} + \frac{\partial^2 \hat{h}_3}{\partial y^2} + (k^2 - k_z^2)\hat{h}_3 &= \frac{k}{2Z_0} H_0^{(1)}(kr\mu)\left(\gamma \int_{-\infty}^{\infty} \delta(z + h)\mathrm{e}^{-ik_z z}\mathrm{d}z - \int_{-\infty}^{\infty} \delta'(z + h)\mathrm{e}^{-ik_z z}\mathrm{d}z\right) \\ &= \frac{k}{2Z_0} H_0^{(1)}(kr\mu)\mathrm{e}^{ik_z h}\left(\frac{ik}{Z_0} - ik_z\right)\end{aligned} \tag{6.27}$$

with $\mu = \mu' + i\mu'' = \sqrt{1 - 1/Z_0^2}$.

2. The second step is a Hankel transform in r defined by

$$H_r[g(r)] = \tilde{f}(k_r) = \int_0^\infty g(r) J_0(k_r r) r \mathrm{d}r \tag{6.28}$$

which gives

$$\left(k^2 - k_r^2 - k_z^2\right)\tilde{\hat{h}}_3 = \frac{k}{2Z_0} i e^{ik_z h}\left(\frac{k}{Z_0} - k_z\right)\int_0^\infty H_0^{(1)}(kr\mu)J_0(k_r r)r\mathrm{d}r \qquad (6.29)$$

or

$$\tilde{\hat{h}}_3 = \frac{k}{2Z_0}\frac{ie^{ik_z h}}{\left(k^2 - k_r^2 - k_z^2\right)}\left(\frac{k}{Z_0} - k_z\right)\int_0^\infty H_0^{(1)}(kr\mu)J_0(k_r r)r\mathrm{d}r. \qquad (6.30)$$

Again, by choosing the branch of the square root in the argument of the Hankel function such that $\mu'' = \operatorname{Im}\mu = \operatorname{Im}\sqrt{1 - 1/Z_0^2} > 0$, the Hankel transform exists. This can be seen by considering the asymptotic expansion of the Hankel function for large arguments r (see [Abr72], p. 364, Eq. 9.2.3)

$$H_0^{(1)}(kr\mu) \approx \sqrt{\frac{2}{\pi kr\mu}}e^{i(kr(\mu'+i\mu'')-\pi/4)}.$$

We find that

$$I = \int_0^\infty H_0^{(1)}(kr\mu)J_0(k_r r)r\mathrm{d}r = \frac{2i}{\pi}\frac{1}{(k\mu)^2 - k_r^2}. \qquad (6.31)$$

This can be easily seen by considering the auxiliary problem (see [Duf01], S. 277)

$$\frac{\mathrm{d}^2 y}{\mathrm{d}r^2} + \frac{1}{r}\frac{\mathrm{d}y}{\mathrm{d}r} + (k\mu)^2 y = -\delta(r)/(2\pi r) \text{ or } H_r[y] = \tilde{y} = \frac{-1}{(k\mu)^2 - k_r^2}\frac{1}{2\pi} \qquad (6.32)$$

with known solution

$$y(r) = \frac{i}{4}H_0^{(1)}(k\mu r). \qquad (6.33)$$

It follows that

$$H_r\left[\frac{i}{4}H_0^{(1)}(k\mu r)\right] = \frac{-1}{(k\mu)^2 - k_r^2}\frac{1}{2\pi} = \frac{i}{4}I \text{ or } I = \frac{2i}{\pi}\frac{1}{(k\mu)^2 - k_r^2}, \qquad (6.34)$$

which confirms (6.31). Therefore,

$$\tilde{\hat{h}}_3 = \frac{\gamma}{\pi}\frac{ie^{ik_z h}}{\left(k^2 - k_r^2 - k_z^2\right)}\left(\frac{k}{Z_0} - k_z\right)\frac{1}{(k\mu)^2 - k_r^2} = w\frac{1}{k_r^2 - \alpha^2}\frac{1}{k_r^2 - \beta^2}, \qquad (6.35)$$

where we have used the abbreviations

$$\alpha^2 = k^2 - k_z^2,\ \beta^2 = (k\mu)^2,\ w = \frac{\gamma}{\pi} i e^{ik_z h}(k/Z_0 - k_z). \tag{6.36}$$

By partial fraction decomposition, we obtain

$$\tilde{\hat{h}}_3 = w\left(\frac{1}{k_r^2 - \alpha^2} - \frac{1}{k_r^2 - \beta^2}\right)\frac{1}{\alpha^2 - \beta^2}. \tag{6.37}$$

3. The third step is the inverse Hankel transform in r. Again by using the identity (6.34), we find that

$$\hat{h}_3 = H_r^{-1}\left[\tilde{\hat{h}}_3\right] = w\left(\frac{i}{4}H_0^{(1)}(\alpha r) - \frac{i}{4}H_0^{(1)}(\beta r)\right)\frac{1}{\alpha^2 - \beta^2} \tag{6.38}$$

or equivalently by substituting back the original quantities defined in (6.36)

$$\hat{h}_3 = \frac{\gamma}{\pi} i e^{ik_z h}(k/Z_0 - k_z)\left(\frac{i}{4}H_0^{(1)}(k\mu r) - \frac{i}{4}H_0^{(1)}\left(r\sqrt{k^2 - k_z^2}\right)\right)\frac{-1}{(k/Z_0)^2 - k_z^2}. \tag{6.39}$$

4. The final step is the inverse Fourier transform in z, which yields

$$h_3 = -\frac{i\gamma}{\pi}\int_{-\infty}^{\infty}\frac{1}{k/Z_0 + k_z}\left(\frac{i}{4}H_0^{(1)}(k\mu r) - \frac{i}{4}H_0^{(1)}\left(r\sqrt{k^2 - k_z^2}\right)\right)e^{ik_z(z+h)}\mathrm{d}k_z \tag{6.40}$$

and finally

$$\begin{aligned} G_3(x, y, z) = {} & \frac{e^{ik\sqrt{x^2+y^2+(z-h)^2}}}{4\pi\sqrt{x^2 + y^2 + (z-h)^2}} + \frac{e^{ik\sqrt{x^2+y^2+(z+h)^2}}}{4\pi\sqrt{x^2 + y^2 + (z+h)^2}} \\ & - \frac{i\gamma}{\pi}\int_{-\infty}^{\infty}\frac{1}{k/Z_0 + k_z}\left(\frac{i}{4}H_0^{(1)}(k\mu r) - \frac{i}{4}H_0^{(1)}\left(r\sqrt{k^2 - k_z^2}\right)\right)e^{ik_z(z+h)}\mathrm{d}k_z. \end{aligned} \tag{6.41}$$

The correction terms h_2, h_3 of the solutions (6.23) for 2D and (6.41) for 3D, respectively, can be written in a unified manner (cp. Eq. (5.89))

$$\begin{aligned} h_m = {} & \frac{-i\gamma}{\pi}\int_{-\infty}^{\infty}\frac{1}{k/Z_0 + k_z}\left(g_{m-1}\left(x, (y); k\sqrt{1 - 1/Z_0^2}\right)\right. \\ & \left. - g_{m-1}\left(x, (y); k\sqrt{1 - k_z^2/k^2}\right)\right)e^{ik_z(z+h)}dk_z \end{aligned} \tag{6.42}$$

for $m = 2, 3$. (Clearly, for $m = 2$ we have no dependence of h_2 on y, which is indicated by the brackets in Eq. (6.42)). Here, g_2 and g_1 are the free-field Green's function (5.80) and (5.82).

6.3 Solutions Represented as an Integral Over Horizontal Plane Waves

We recall that the correction term of the driving sources for the complex monopole in the 1D–3D half-space can be written as (see Eqs. (5.67), (B5a), and (C5a))

$$S_1 = \delta(z-h) + \delta(z+h) - \frac{2\gamma}{k^2+\gamma^2}\{\gamma\delta(z_h) - \delta'(z_h)\}, \tag{6.43}$$

$$S_2 = \delta(x, z-h) + \delta(x, z+h) - F_x^{-1}\left[\frac{2\gamma}{\kappa_2^2+\gamma^2}\right]\{\gamma\delta(z_h) - \delta'(z_h)\}, \tag{6.44}$$

$$S_3 = \delta(x, y, z-h) + \delta(x, y, z+h) - \frac{1}{2\pi}H_r^{-1}\left[\frac{2\gamma}{\kappa_3^2+\gamma^2}\right]\{\gamma\delta(z_h) - \delta'(z_h)\} \tag{6.45}$$

with

$$\kappa_2 = \sqrt{k^2 - k_x^2},\ \kappa_3 = \sqrt{k^2 - k_r^2} \text{ and } r = \sqrt{x^2 + y^2}.$$

Equations (5.87) and (5.88) were obtained by carrying out the inverse Fourier or Hankel transforms explicitly. However, this was not done in Eqs. (6.44) and (6.45), so that we can perform a Fourier transform in x or a Hankel transform in r very easily and get immediately for the third terms on the RHS of (6.44) and (6.45), respectively

$$\tilde{\chi}_2^{(2)} := F_x\left[\chi_2^{(2)}\right] = -\left[\frac{2\gamma}{\kappa_2^2+\gamma^2}\right]\{\gamma\delta(z_h) - \delta'(z_h)\} \tag{6.46}$$

or

$$\tilde{\chi}_3^{(2)} := H_r\left[\chi_3^{(2)}\right] = -\frac{1}{2\pi}\left[\frac{2\gamma}{\kappa_3^2+\gamma^2}\right]\{\gamma\delta(z_h) - \delta'(z_h)\}. \tag{6.47}$$

After these preparations, we consider the 2D and the 3D case separately:

6.3.1 Solutions Represented as an Integral Over Horizontal Plane Waves in 2D

In order to obtain the correction term of the solution for the rigid case, we have to solve the second order differential equation (cp. Eq. (6.7) for the Fourier transform of $h_2(x, z)$)

$$\frac{\mathrm{d}^2\tilde{h}_2(z)}{\mathrm{d}z^2} + \kappa_2^2\tilde{h}_2(z) = -\chi_2^{(2)} = -\left[\frac{-2\gamma}{\kappa_2^2+\gamma^2}\right]\{\gamma\delta(z_h) - \delta'(z_h)\} \tag{6.48}$$

with well-known solution

$$\tilde{h}_2(z) = \frac{-2\gamma}{\kappa_2^2+\gamma^2}\frac{i}{2\kappa_2}\{\gamma\mathrm{e}^{i\kappa_2 z_h} - i\kappa_2\gamma\mathrm{e}^{i\kappa_2 z_h}\} = \frac{-ik}{\kappa_2(k+Z_0\kappa_2)}\mathrm{e}^{i\kappa_2 z_h} \quad \text{for } z_h > 0. \tag{6.49}$$

Finally, we obtain by inverse Fourier transform in x

$$\begin{aligned} h_2(x,z) &= \frac{-ik}{2\pi}\int\limits_{-\infty}^{\infty}\frac{\mathrm{e}^{i\sqrt{k^2-k_x^2}|z_h|}\mathrm{e}^{ik_x x}}{\sqrt{k^2-k_x^2}\left(k+Z_0\sqrt{k^2-k_x^2}\right)}\mathrm{d}k_x \\ &= \frac{-ik}{\pi}\int\limits_{0}^{\infty}\frac{\mathrm{e}^{i\sqrt{k^2-k_x^2}|z_h|}\cos(k_x x)}{\sqrt{k^2-k_x^2}\left(k+Z_0\sqrt{k^2-k_x^2}\right)}\mathrm{d}k_x. \end{aligned} \tag{6.50}$$

This formula is exactly the same as Chandler-Wilde and Hothersall has found in [Cha95], Eq. (13). It can be transformed into a more convenient trigonometric form and into representations suitable for numerical evaluations as shown in [Cha95]. Also, Mechel has used an integral representation of the Hankel function consisting of plane waves (see [Mec00], Eq. (9)) and has found similar trigonometric expressions for the scattered pressure.

6.3.2 Solutions Represented as an Integral Over Horizontal Plane Waves in 3D

Proceeding just as in the same way as in 2D with the only difference that the Hankel transform has to be used instead of the Fourier transform in x, we obtain

$$h_3(x,y,z) = \frac{-ik}{2\pi}\int\limits_{0}^{\infty}\frac{\mathrm{e}^{i\sqrt{k^2-k_r^2}|z_h|}J_0(k_r r)k_r\mathrm{d}k_r}{\sqrt{k^2-k_r^2}\left(k+Z_0\sqrt{k^2-k_r^2}\right)}. \tag{6.51}$$

In order to compare this solution with the expression given by Wenzel [Wen74], we substitute

$$\sqrt{k^2 - k_r^2} = i\sqrt{k_r^2 - k^2} =: im$$

which gives

$$h_3(x, y, z) = \frac{-ik}{2\pi} \int_0^\infty \frac{\mathrm{e}^{-m|z_h|} J_0(k_r r) k_r \mathrm{d}k_r}{im(k + Z_0 im)} = \frac{-\gamma}{2\pi} \int_0^\infty \frac{\mathrm{e}^{-m|z_h|} J_0(k_r r) k_r \mathrm{d}k_r}{m(\gamma - m)}. \quad (6.52)$$

Expression (6.52) is identical with Eq. (10) of the work of Wenzel [Wen74], where the different minus signs are caused by the different definitions of the RHS of the Helmholtz equation.

In summary, the inhomogeneous Helmholtz equation for the impedance half-space problem makes it possible to derive several different but equivalent representations for the solution in 2D and in 3D, which can be found in part in the literature.

In the following Chaps. 7–11, *the imaginary unit is denoted by* $j = \sqrt{-1}$ *and the time convention* $\mathrm{e}^{+i\omega t}$ *is used (except in the section about the EIT).*

Chapter 7
Branch Cuts of the Square Root with Complex Argument

7.1 Complex Source Position

A complex source position for the time convention $e^{j\omega t}$ is given by

$$\vec{r}_q^{\,c} = \vec{r}_q - j\beta\hat{u}^c, \tag{7.1}$$

where $\vec{r}_q$ is the real part and $\beta\hat{u}^c$ the imaginary part with β a positive real scalar and $\hat{u}^c$ a unitary real vector. The complex distance function between a real field point and the complex source position is the complex square root R^c

$$R^c = \sqrt{\left(\vec{r} - \vec{r}_q^{\,c}\right) \cdot \left(\vec{r} - \vec{r}_q^{\,c}\right)}. \tag{7.2}$$

If θ is the angle between the vectors $\vec{r} - \vec{r}_q$ and $\hat{u}$ and $R_q = \left|\vec{r} - \vec{r}_q\right|$

$$R^c = \sqrt{R_q^2 - \beta^2 + j2\beta R_q \cos\theta}. \tag{7.3}$$

Hence, R^c has **axial symmetry** respect to the axis defined by the unit vector $\hat{u}^c$. By writing $R_q^2 = R_q^2\cos^2\theta + R_q^2\sin^2\theta$ we can rewrite (7.3) as

$$R^c = \sqrt{R_q^2\sin^2\theta - \left(\beta - jR_q\cos\theta\right)^2} \tag{7.4}$$

and rearrange it as

$$R^c = \sqrt{\left(R_q\sin\theta - jR_q\cos\theta + \beta\right)\left(R_q\sin\theta + jR_q\cos\theta - \beta\right)}, \tag{7.5}$$

or more conveniently as

M. Ochmann and R. Piscoya, *Theory and Application of Acoustic Sources Using Complex Analysis*, https://doi.org/10.1007/978-981-33-6040-2_7

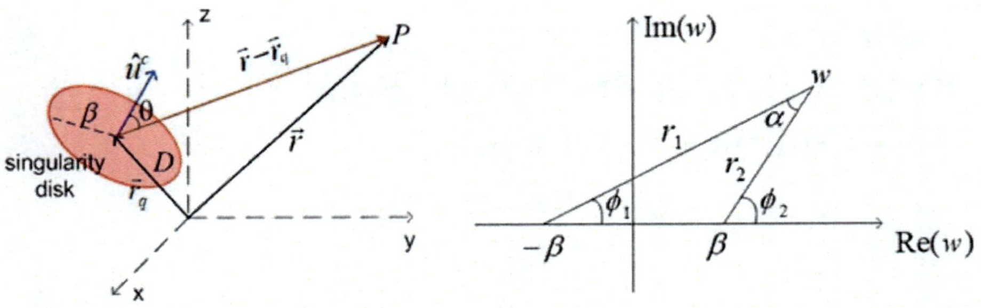

Fig. 7.1 Parameters of complex source in real space (left) and for computing the complex square root in w-plane (right)

$$R^c = \sqrt{\left(R_q \sin\theta + jR_q\cos\theta + \beta\right)^* \left(R_q \sin\theta + jR_q\cos\theta - \beta\right)}, \tag{7.6}$$

where $()^*$ means complex conjugate. By introducing the complex variable $w = R_q \sin\theta + jR_q\cos\theta$ we have

$$R^c = \sqrt{(w+\beta)^*(w-\beta)}. \tag{7.7}$$

In the physical space, Re(w) corresponds to the radial coordinate perpendicular to $\hat{u}^c$ and Im(w) to the axial coordinate parallel to $\hat{u}^c$.

By denoting

$$w + \beta = r_1 e^{j\phi_1},\ w - \beta = r_2 e^{j\phi_2} \tag{7.8}$$

with r_1, r_2, ϕ_1 and ϕ_2 as shown in Fig. 7.1, R^c is defined as

$$R^c = \sqrt{r_1 r_2} e^{j[(\phi_2-\phi_1)/2+n\pi]}, \quad n = \{0, 1\} \tag{7.9}$$

One (principal) branch of (7.9) needs to be chosen. It is customary, though arbitrary, to take as principal branch the one with $n = 0$. The value of R^c will depend on the values of ϕ_1 and ϕ_2. Additionally, Re(R^c) and/or Im(R^c) will be discontinuous at some curve in the w-plane due to the different periods of $e^{j(\phi_2-\phi_1)}$ and $e^{j(\phi_2-\phi_1)/2}$. In order to have a well-behaved single-valued function, it is also customary to introduce a branch cut along the curve of discontinuity and agree not to cross the cut in the w-plane. The branch cuts usually join two branch points or one branch point to the infinity. The branch points are found by setting $R^c = 0$. The branch points of R^c are $w = -\beta$ and $w = +\beta$.

From (7.9), $\arg(R^c) = \frac{\phi_2-\phi_1}{2}$ and from Fig. 7.1 we find that,

$$\arg(R^c) = \frac{\phi_2-\phi_1}{2} = \frac{\alpha}{2}, \tag{7.10}$$

and to determine $\arg(R^c)$, the values of ϕ_1 and ϕ_2 need to be specified. We show three possible choices for the values of ϕ_1 and ϕ_2 (branches).

7.1.1 "Source" Branch

In order to have sound waves moving away from the source, $\text{Re}(R^c) > 0$. This condition is fulfilled by requiring that

$$-\frac{\pi}{2} < \arg(R^c) < \frac{\pi}{2}. \tag{7.11}$$

According to (7.11), $\text{Im}(R^c)$ will be discontinuous and the curve of discontinuity can be obtained by setting $\arg(R^c) = \pm\pi/2$ (or equivalently $\text{Re}(R^c) = 0$). By defining $w = u + jv$,

$$\left(R^c\right)^2 = u^2 + v^2 - \beta^2 + j2\beta v, \tag{7.12}$$

and $\text{Re}(R^c) = 0$ if

$$v = 0 \text{ and } u^2 + v^2 - \beta^2 < 0 \rightarrow -\beta < u < \beta. \tag{7.13}$$

In the w-plane, the proper branch cut will be given by (7.13) and joins both branch points $w = -\beta$ and $w = +\beta$. To obtain the desired branch cut, the range of both angles ϕ_1 and ϕ_2 should be

$$\phi_1 \in [-\pi, \pi], \phi_2 \in [-\pi, \pi]. \tag{7.14}$$

The choice (7.14) is done by the introduction of two branch cuts, each joining a branch point with $-\infty$. The combination of both branch cuts produce an effective branch cut for the square root at $u \in [-\beta, \beta]$ as intended (see Fig. 7.2). This can be corroborated by analyzing $\arg(R^c)$ at a set of points very close to the u-axis (Table 7.1).

Since geometrically $\phi_2 - \phi_1 = \alpha$ and $|\phi_2| \geq |\phi_1|$, it can be seen that:

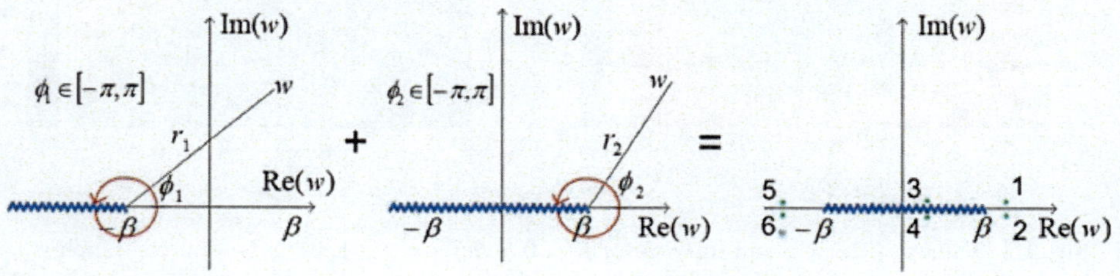

Fig. 7.2 Illustration of the source branch cut

Table 7.1 Values of $\arg(R^c)$ close to the source branch cut

Point	ϕ_1	ϕ_2	$\arg(R^c)$
1	0	0	0
2	0	0	0
3	0	π	$\pi/2$
4	0	$-\pi$	$-\pi/2$
5	π	π	0
6	$-\pi$	$-\pi$	0

For $v \geq 0$, ϕ_1 and ϕ_2 are both positive so that $\alpha \in [0, \pi]$, i.e., $\mathrm{Im}(R^c) > 0$ with

$$\alpha\left(v = 0^+\right) = \begin{cases} \pi & |u| < \beta \\ 0 & |u| > \beta \end{cases}. \tag{7.15}$$

For $v \leq 0$, ϕ_1 and ϕ_2 are both negative and $\alpha \in [-\pi, 0]$, i.e., $\mathrm{Im}(R^c) < 0$ with

$$\alpha\left(v = 0^-\right) = \begin{cases} -\pi & |u| < \beta \\ 0 & |u| > \beta \end{cases}. \tag{7.16}$$

Therefore, the choice (7.14) provides us with the desired condition $\mathrm{Re}(R^c) > 0$ (Fig. 7.3).

In the physical space, this branch cut corresponds to the singularity disk of radius β that lie on the plane P

$$P : \left(\vec{r} - \vec{r}_q\right) \cdot \hat{u} = 0. \tag{7.17}$$

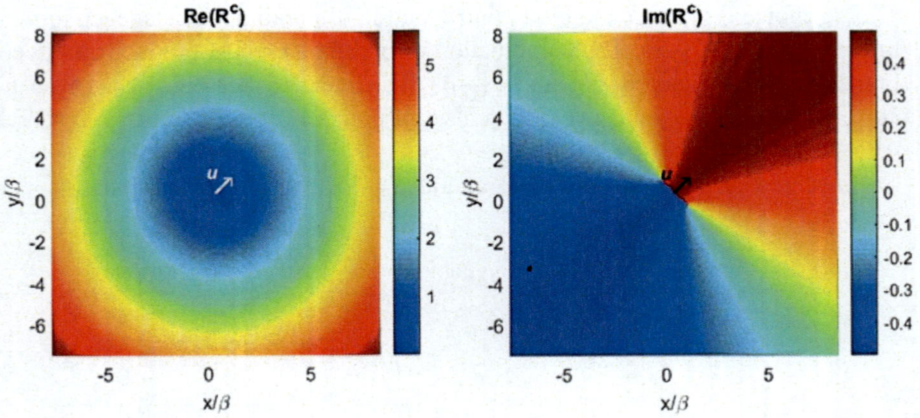

Fig. 7.3 Values of $\mathrm{Re}(R^c)$ and $\mathrm{Im}(R^c)$ for $\beta = 0.5$ and $\hat{u} = (1, 1, 0)/\sqrt{2}$—source branch

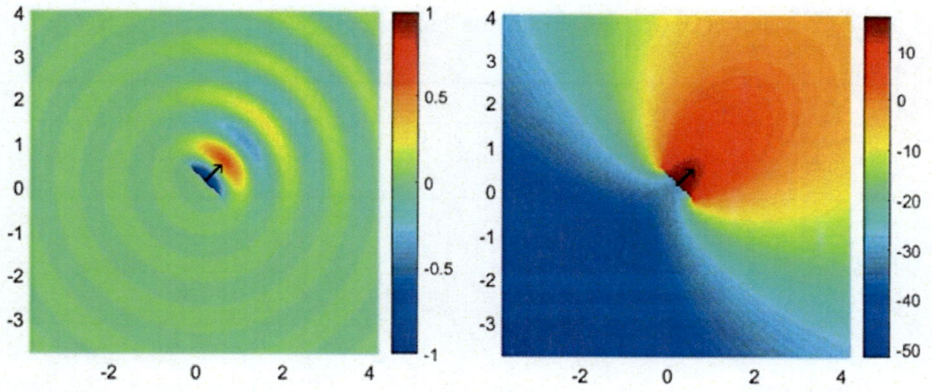

Fig. 7.4 Sound radiation of the complex monopole (source branch); left: pressure; right: sound level

So, $\text{Im}(R^c) > 0$ above the plane P and $\text{Im}(R^c) < 0$ below the plane P with $\text{Im}(R^c) = 0$ on the plane, except within the singularity disk where $\arg(R^c)$ changes from $\pi/2$ above the disk to $-\pi/2$ below it.

Since $\text{Re}(R^c) > 0$ everywhere, the sound waves propagate in all directions, but the amplitude of the waves is bigger in the region $\text{Im}(R^c) > 0$ than in the region $\text{Im}(R^c) < 0$ (Fig. 7.4).

7.1.2 "Beam" Branch Cut

A different selection of the values of ϕ_1 and ϕ_2 can be given by

$$\phi_1 \in [-\pi, \pi], \phi_2 \in [0, 2\pi]. \tag{7.18}$$

The choice (7.18) is obtained by introducing the branch cuts that joins the branch point $-\beta$ with $-\infty$ and the branch point $+\beta$ with $+\infty$ (Fig. 7.5).

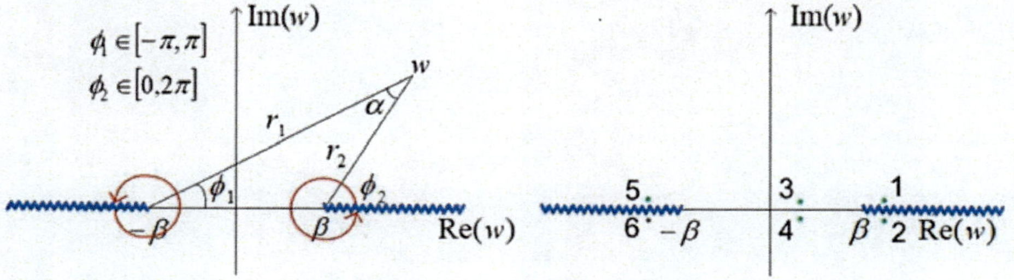

Fig. 7.5 Illustration of the beam branch cut

Table 7.2 Values of $\arg(R^c)$ close to the beam branch cut

Point	ϕ_1	ϕ_2	$\arg(R^c)$
1	0	0	0
2	0	2π	π
3	0	π	$\pi/2$
4	0	π	$\pi/2$
5	π	π	0
6	$-\pi$	π	π

These branch cuts yield a discontinuity of $\mathrm{Re}(R^c)$ for $|u| > \beta$. This can be corroborated by analyzing $\arg(R^c)$ at a set of points very close to the u-axis (Table 7.2).

Similarly to the source case, for $v \geq 0$, ϕ_1 and ϕ_2 are both positive and have values in the interval $[0, \pi]$ so that $\alpha \in [0, \pi]$, i.e., $\mathrm{Re}(R^c) > 0$ with

$$\alpha\left(v = 0^+\right) = \begin{cases} \pi & |u| < \beta \\ 0 & |u| > \beta \end{cases} \tag{7.19}$$

For $v \leq 0$, ϕ_1 is negative but ϕ_2 is positive, therefore $\alpha \in [\pi, 2\pi]$, i.e., $\mathrm{Re}(R^c) < 0$ with

$$\alpha\left(v = 0^-\right) = \begin{cases} \pi & |u| < \beta \\ 2\pi & |u| > \beta \end{cases}. \tag{7.20}$$

In the physical space, this branch cut corresponds to the region of the plane P outside the singularity ring. So, $\mathrm{Re}(R^c) \geq 0$ above the plane P and $\mathrm{Re}(R^c) \leq 0$ below the plane P. $\arg(R^c)$ changes from 0 slightly above the plane to π slightly below it outside the singularity ring, but $\mathrm{Re}(R^c) = 0$ on the plane inside the singularity ring (Fig. 7.6).

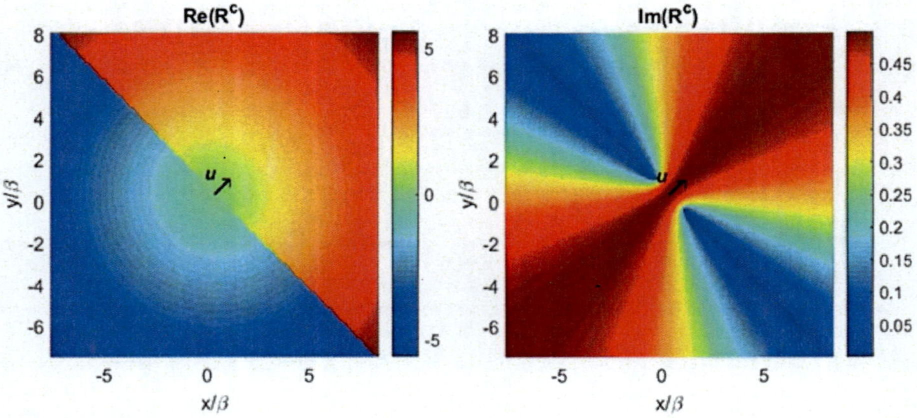

Fig. 7.6 Values of $\mathrm{Re}(R^c)$ and $\mathrm{Im}(R^c)$ for $\beta = 0.5$ and $\hat{u} = (1, 1, 0)/\sqrt{2}$—beam branch

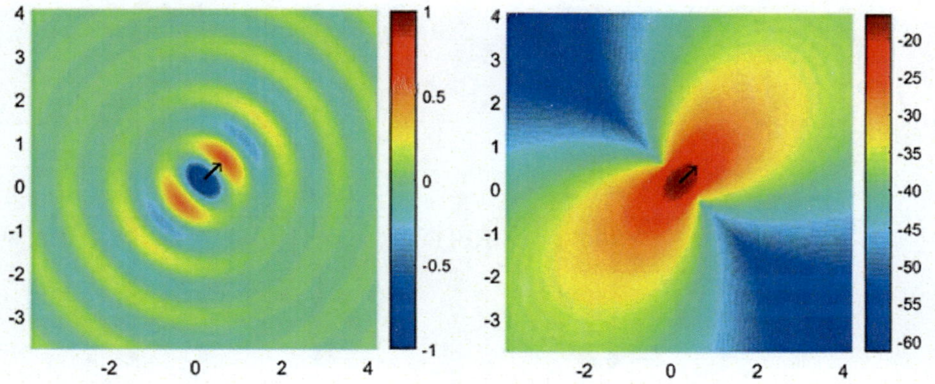

Fig. 7.7 Sound radiation of the complex monopole—beam branch; left: pressure; right: sound level

Because $\mathrm{Re}(R^c)$ changes from negative values below the plane to positive values above it, the sound waves would come from below the plane, converge inside the singularity ring, pass through it and diverge from the singularity ring into the open positive half-space (Fig. 7.7). The radiation condition is fulfilled only above the plane P.

7.1.3 "Open Duct" Branch

Another selection of the values of ϕ_1 and ϕ_2 can be given by

$$\phi_1 = \left[-\frac{\pi}{2}, \frac{3\pi}{2}\right], \phi_2 = \left[-\frac{\pi}{2}, \frac{3\pi}{2}\right]. \tag{7.21}$$

The choice (7.21) is obtained by introducing the branch cuts that join the branch point $-\beta$ with $-\beta - j\infty$ and the branch point $+\beta$ with $\beta - j\infty$. (Fig. 7.8).

These branch cuts imply that $\arg(R^c)$ is discontinuous for $|\mathrm{Re}(w)| = \beta$ and $\mathrm{Im}(w) < 0$. This can be corroborated by analyzing $\arg(R^c)$ at a set of points very close to the branches (Table 7.3).

As in the previous two cases, for $v \geq 0$, ϕ_1 and ϕ_2 are both positive and have values in the interval $[0, \pi]$ so that $\alpha \in [0, \pi]$, i.e., $\mathrm{Re}(R^c) > 0$ with

$$\alpha(v = 0^+) = \begin{cases} \pi & |u| < \beta \\ 0 & |u| > \beta \end{cases}. \tag{7.22}$$

For $v \leq 0$, ϕ_1 is negative for $u > -\beta$ and positive for $u < -\beta$ while ϕ_2 is positive for $u < \beta$, and negative for $u > \beta$ therefore $\alpha \in [-\pi/2, 0]$, i.e., $\mathrm{Re}(R^c) > 0$ for

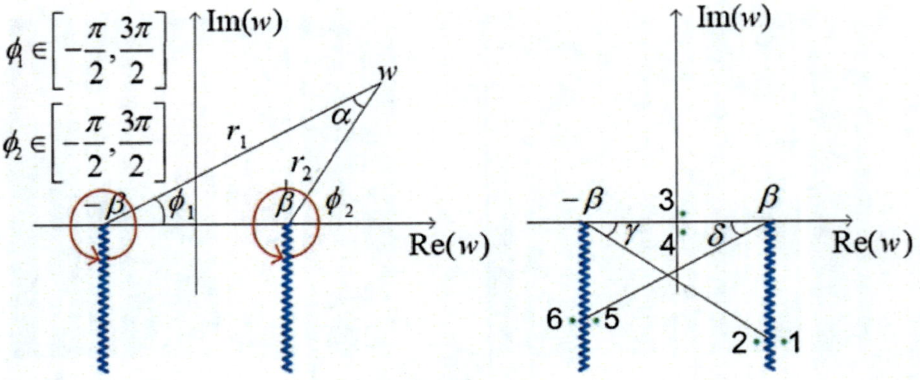

Fig. 7.8 Illustration of the open duct branch cut

Table 7.3 Values of $\arg(R^c)$ close to the beam branch cut

Point	ϕ_1	ϕ_2	$\arg(R^c)$
1	$-\gamma$	$-\pi/2$	$-\pi/4+\gamma/2$
2	$-\gamma$	$3\pi/2$	$3\pi/4+\gamma/2$
3	0	π	$\pi/2$
4	0	π	$\pi/2$
5	$-\pi/2$	$\pi+\delta$	$\delta/2+3\pi/4$
6	$3\pi/2$	$\pi+\delta$	$\delta/2-\pi/4$

$|u|>\beta$ and $\alpha\in[3\pi/2,2\pi]$, i.e., $\mathrm{Re}(R^c)<0$ for $|u|<\beta$ with

$$\alpha\left(v=0^-\right)=\begin{cases}\pi & |u|<\beta\\ 0 & |u|>\beta\end{cases}. \tag{7.23}$$

In the physical space, this branch cut corresponds to the semi-infinite cylindrical surface with radius β and axis parallel to the $\hat{u}$-axis region placed below the plane P. Since $\mathrm{Re}(R^c)<0$ inside the semi-infinite cylinder and $\mathrm{Re}(R^c)>0$ outside (Fig. 7.9), the sound waves would travel along the cylinder toward the opening at the singular ring and then propagate into the open space, similar to the sound radiation of an unflanged semi-infinite duct (Fig. 7.10).

7.2 Sommerfeld Branches

Consider two semi-infinite homogeneous media with densities ρ_1 and ρ_2 and sound speeds c_1 and c_2, respectively, with the interface between the media at $z=0$. The velocity potential due to a point source with volume flow Q_0 at a height h from the

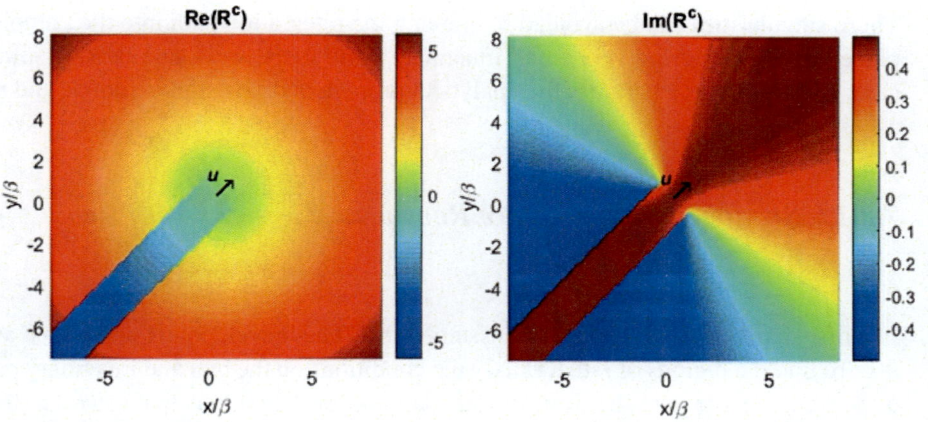

Fig. 7.9 Values of $\mathrm{Re}(R^c)$ and $\mathrm{Im}(R^c)$ for $\beta = 0.5$ and $\hat{u} = (1, 1, 0)/\sqrt{2}$—open duct branch

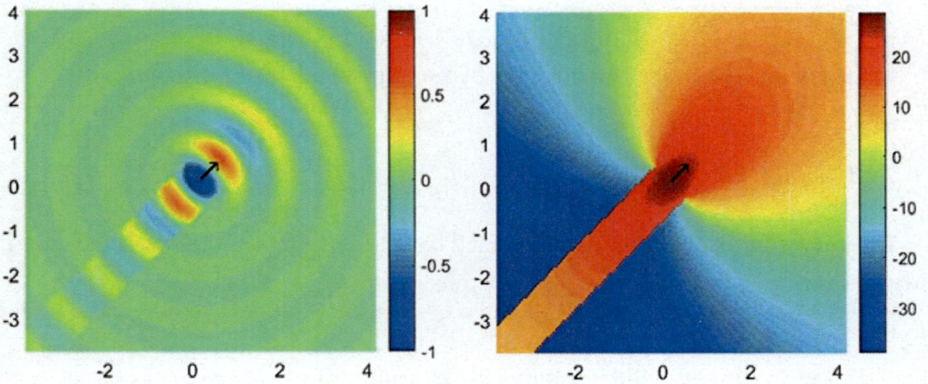

Fig. 7.10 Sound radiation of the complex monopole—open duct branch; left: pressure; right: sound level

interface is given by

$$\phi(x, y, z) = \begin{cases} Q_0 \frac{\mathrm{e}^{-jk_1 R_1}}{4\pi R_1} + \frac{Q_0}{4\pi} \int\limits_0^\infty J_0(\lambda\rho) \mathrm{e}^{-\ell(z+h)} \frac{(\rho_2 \ell - \rho_1 m)}{\rho_2 \ell + \rho_1 m} \lambda \mathrm{d}\lambda & z > 0 \\ \frac{2\rho_1 Q_0}{4\pi} \int\limits_0^\infty J_0(\lambda\rho) \frac{\mathrm{e}^{-\ell h} \mathrm{e}^{mz}}{\rho_2 \ell + \rho_1 m} \lambda \mathrm{d}\lambda & z < 0 \end{cases}, \tag{7.24}$$

where k_1 and k_2 are the wave numbers, $\ell = \sqrt{\lambda^2 - k_1^2}$ and $m = \sqrt{\lambda^2 - k_2^2}$ are propagation constants and R_1 is the distance between source and receiver.

Since the integrands comprise waves propagating along the $+z$ and $-z$, respectively, the imaginary part of the propagation constants must be positive. Additionally, to have converging integrals, the real part of the propagation constants must also be positive. For real wave numbers k_1 and k_2, the integrands will have singularities.

These singularities can be avoided by moving the integration path into the complex plane or by considering very small imaginary parts in the wave numbers. In either case, it is necessary to discuss the analytical properties of the square roots ℓ and m.

7.2.1 *Branches of the Square Root* $\mu = \sqrt{\lambda^2 - k^2}$ *with Real* k

In a first step, the wave number k is assumed real. The branch points will be $-k$ and $+k$. To find the regions of λ that satisfy the conditions on the real and imaginary part of μ, we can (i) require that $\mathrm{Re}(\mu) > 0$ and look for the regions that $\mathrm{Im}(\mu) > 0$ or (ii) require that $\mathrm{Im}(\mu) > 0$ and look for the regions that $\mathrm{Re}(\mu) > 0$.

By defining $\lambda = u + jv$

$$\mu^2 = u^2 - v^2 - k^2 + j2uv, \tag{7.25}$$

we can define the proper branch cuts by finding the curves of discontinuity.

7.2.1.1 Re(μ) > 0

The curves of discontinuity are obtained by setting $\mathrm{Re}(\mu) = 0$. That is the case if

$$uv = 0 \text{ and } u^2 - v^2 - k^2 < 0. \tag{7.26}$$

If $u = 0$, $k^2 + v^2$ will be positive for all values of v.

If $v = 0 \rightarrow u^2 - k^2 < 0$, hence, the curve of discontinuity will be $|u| < k$.

The proper branch cut can be constructed by combining the two branches,

$$\begin{aligned} B_{11} &= \{-k < u < 0 \cap v = 0\} \cup \{0 < v < \infty \cap u = 0\} \\ B_{12} &= \{-\infty < v < 0 \cap u = 0\} \cup \{0 < u < k \cap v = 0\} \\ B_1 &= B_{11} + B_{12} \end{aligned} \tag{7.27}$$

or the two branches (Fig. 7.11)

$$\begin{aligned} B_{21} &= \{-k < u < 0 \cap v = 0\} \cup \{-\infty < v < 0 \cap u = 0\} \\ B_{22} &= \{0 < v < \infty \cap u = 0\} \cup \{0 < u < k \cap v = 0\} \\ B_2 &= B_{21} + B_{22}. \end{aligned} \tag{7.28}$$

$\mathrm{Im}(\mu)$ will be discontinuous on the branch cut as can be corroborated by analyzing $\arg(\mu)$ at a set of points very close to the branches (Table 7.4).

By denoting

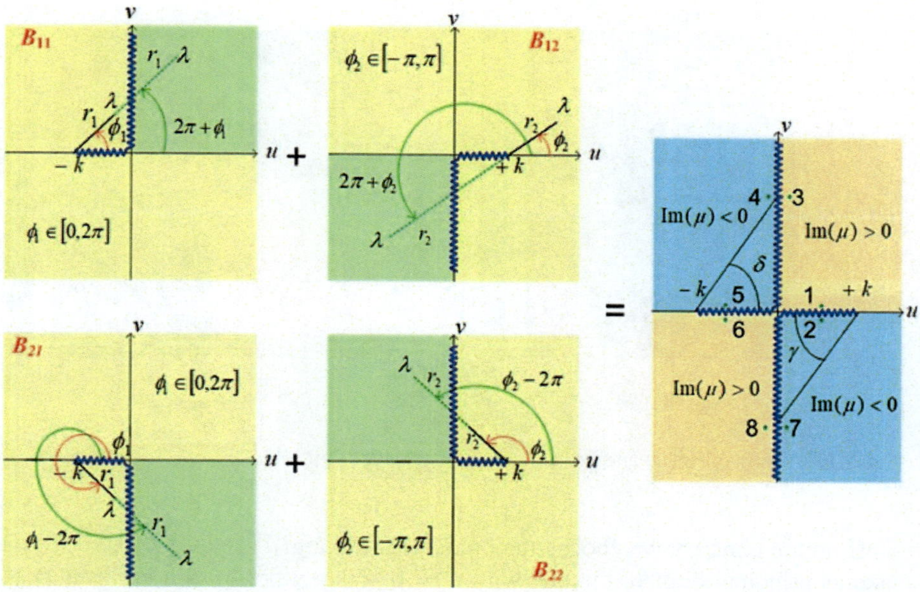

Fig. 7.11 Illustration of the branch cuts for $\mathrm{Re}(\mu) = 0$

Table 7.4 Values of $\arg(\mu)$ close to the branch cut

Point	B_1			B_2		
	ϕ_1	ϕ_2	$\arg(\mu)$	ϕ_1	ϕ_2	$\arg(\mu)$
1	2π	π	$3\pi/2$	0	π	$\pi/2$
2	2π	$-\pi$	$\pi/2$	0	$-\pi$	$-\pi/2$
3	$2\pi + \delta$	$\pi - \delta$	$3\pi/2$	δ	$\pi - \delta$	$\pi/2$
4	δ	$\pi - \delta$	$\pi/2$	δ	$-\pi - \delta$	$-\pi/2$
5	0	π	$\pi/2$	0	$-\pi$	$-\pi/2$
6	2π	π	$3\pi/2$	2π	$-\pi$	$\pi/2$
7	$2\pi - \gamma$	$\gamma - \pi$	$\pi/2$	$-\gamma$	$\gamma - \pi$	$-\pi/2$
8	$2\pi - \gamma$	$\pi + \gamma$	$3\pi/2$	$2\pi - \gamma$	$\gamma - \pi$	$\pi/2$

$$\lambda + k = r_1 \mathrm{e}^{j\phi_1}, \lambda - k = r_2 \mathrm{e}^{j\phi_2} \tag{7.29}$$

μ is defined as

$$\mu = \sqrt{r_1 r_2}\mathrm{e}^{j[(\phi_1+\phi_2)/2+n\pi]}, \quad n = \{0, 1\}. \tag{7.30}$$

The values of ϕ_1, ϕ_2 and n will be determined by the specific choice of the branch cuts (7.27) or (7.28). If we choose the branch cut B_1, $\arg(\mu) \in \left[\frac{\pi}{2}, \frac{3\pi}{2}\right]$ and since we require that $\mathrm{Re}(\mu) > 0$, we must take as the principal branch the one with $n = 1$, so that $\arg(\mu) \in \left[\frac{3\pi}{2}, \frac{5\pi}{2}\right]$.

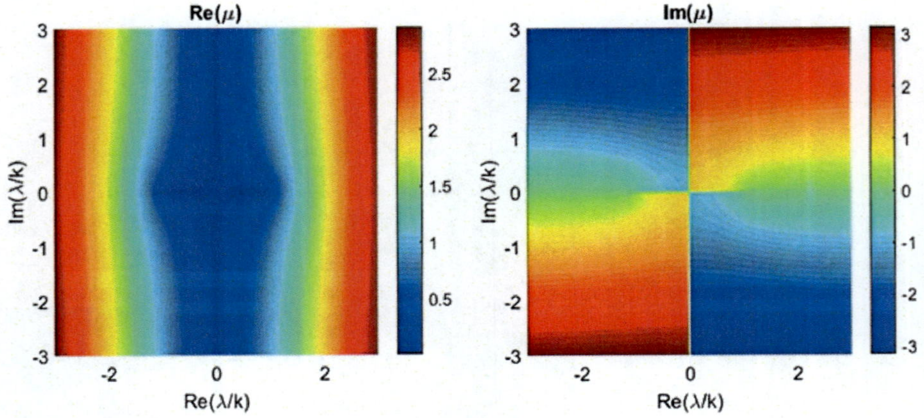

Fig. 7.12 Values of Re(μ) and Im(μ) for the branch cut of Fig. 7.11

If on the contrary we choose the branch cut B_2, $\arg(\mu) \in \left[-\frac{\pi}{2}, \frac{\pi}{2}\right]$, so we must take as principal branch, the one with $n = 0$. With either branch cut, $\text{Im}(\mu) > 0$ in the first and third quadrants and $\text{Im}(\mu) < 0$ in the second and fourth quadrants (Fig. 7.12).

7.2.1.2 Im(μ) > 0

The curves of discontinuity are obtained by setting $\text{Im}(\mu) = 0$. That is the case if

$$uv = 0 \text{ and } u^2 - v^2 - k^2 > 0. \tag{7.31}$$

If $u = 0$, $k^2 + v^2$ will never be negative.

If $v = 0 \rightarrow u^2 - k^2 > 0$, hence, the curve of discontinuity will be $|u| > k$.

The proper branch cut will be similar to the "beam branch" of Fig. 7.5. The branch cuts yield a discontinuity of Re(μ) for $|u| > k$. The change in $\arg(\mu)$ at the same set of points very close to the u-axis of Fig. 7.5 is given in Table 7.5.

Table 7.5 Values of $\arg(\mu)$ close to the branch cut of Fig. 7.5

Point	ϕ_1	ϕ_2	$\arg(\mu)$
1	0	0	0
2	0	2π	π
3	0	π	$\pi/2$
4	0	π	$\pi/2$
5	π	π	π
6	$-\pi$	π	0

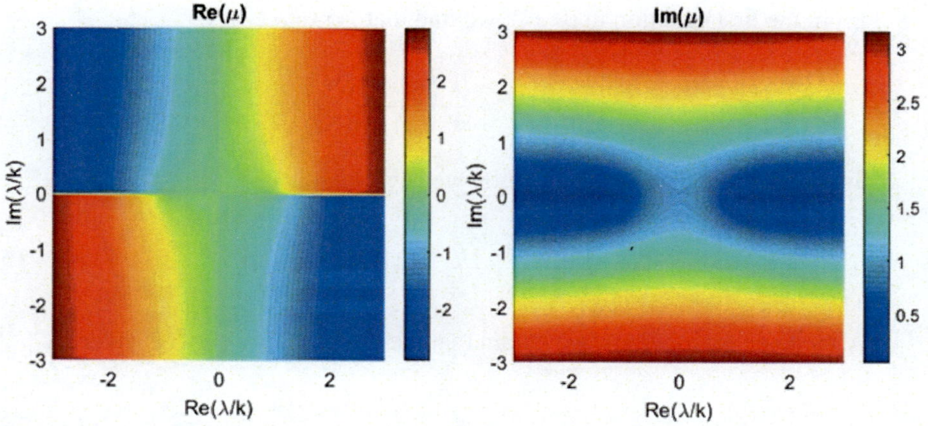

Fig. 7.13 Values of Re(μ) and Im(μ) for the branch cut of Fig. 7.5

Since we require that $\mathrm{Im}(\mu) > 0$, for this case, we must take as the principal branch the one with $n = 0$, so that $\arg(\mu) \in [0, \pi]$. With the chosen branch cut, $\mathrm{Re}(\mu) > 0$ in the first and third quadrants and $\mathrm{Re}(\mu) < 0$ in the second and fourth quadrants (Fig. 7.13).

7.2.2 *Branches of the Square Root* $\mu = \sqrt{\lambda^2 - k^2}$ *with Complex* k

Since we are dealing with sound waves propagation, we will require that $\mathrm{Re}(k) > 0$, hence we will write k as

$$k = |k|e^{j\beta}, -\frac{\pi}{2} < \beta < \frac{\pi}{2} \tag{7.32}$$

and μ^2 is now given by

$$\mu^2 = u^2 - v^2 + j2uv - |k|^2(\cos 2\beta + j\sin 2\beta) \tag{7.33}$$

We find the curves of discontinuity.

7.2.2.1 Re(μ) > 0

The curves of discontinuity are obtained by setting $\mathrm{Re}(\mu) = 0$. That is the case if

$$2uv - |k|^2 \sin 2\beta = 0 \text{ and } u^2 - v^2 - k^2\cos 2\beta < 0. \tag{7.34}$$

From the first equation in (7.34), we find that

$$v = \frac{|k|^2 \sin 2\beta}{2u} \tag{7.35}$$

and replacing into the second equation we have

$$u^2 - \frac{|k|^4 \sin^2 2\beta}{4u^2} - |k|^2 \cos 2\beta < 0. \tag{7.36}$$

After some calculations, we obtain that the discontinuity curve is given by

$$|u| < |k| \cos \beta, \, v = \frac{|k|^2 \sin 2\beta}{2u}. \tag{7.37}$$

We can parameterize the curve and rewrite it as

$$u = t|k| \cos \beta, \, v = \frac{|k| \sin \beta}{t}, \, t \in [-1, 1]. \tag{7.38}$$

According to (7.32), $\cos \beta$ is always positive, but $\sin \beta$ can be positive or negative. Therefore, for $0 < \beta < \pi/2$, the curves will occupy the first and third quadrants, while for $-\pi/2 < \beta < 0$, the second and fourth quadrants. The curve is discontinuous at $t = 0$.

$$t = 0^+, \lambda \to \text{sgn}(\sin \beta) j\infty \text{ and } t = 0^-, \lambda \to -\text{sgn}(\sin \beta) j\infty. \tag{7.39}$$

Figure 7.14 shows the proper branch cuts for $0 < \beta < \pi/2$ and illustrates the angles ϕ_1 and ϕ_2 which require to take $n = 0$. The regions where $\text{Im}(\mu) > 0$ and $\text{Im}(\mu) < 0$ are indicated as well. For the case $-\pi/2 < \beta < 0$, the branch cuts are images of the ones from Fig. 7.14 respect to the v-axis and the angles ϕ_1 and ϕ_2 are

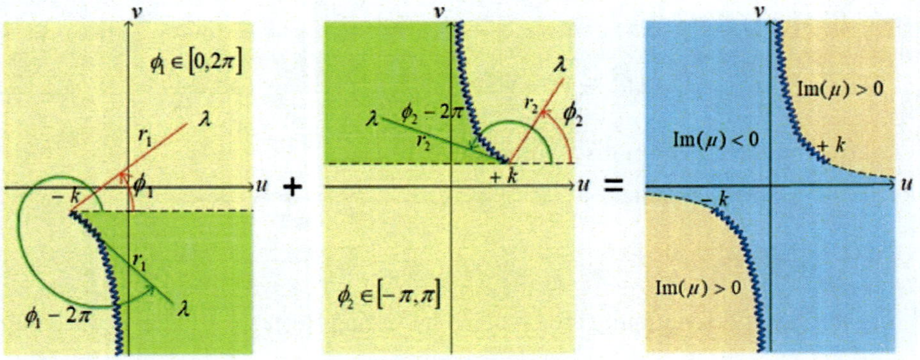

Fig. 7.14 Illustration of the branch cuts for $\text{Re}(\mu) = 0$ and k complex

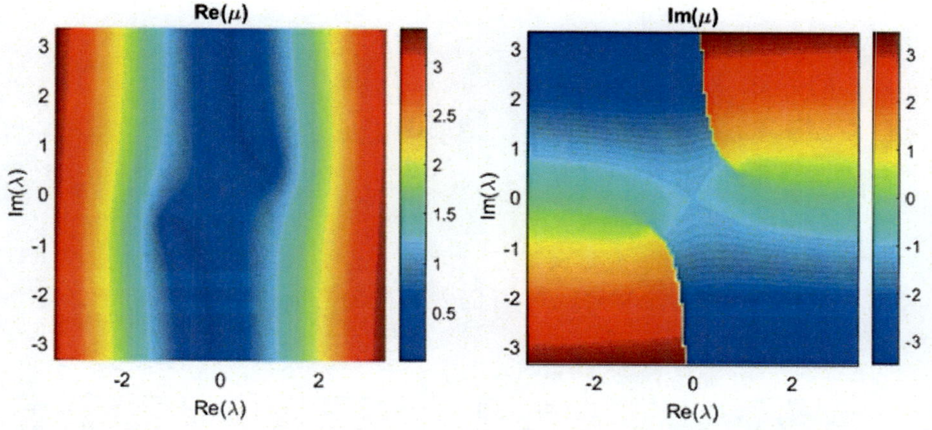

Fig. 7.15 Values of Re(μ) and Im(μ) for the branch cut of Fig. 7.14

similarly defined as those from Fig. 7.11 (branch B_1). For that case, the principal branch is the one with $n = 1$.

Figure 7.15 shows the real and imaginary parts of $\sqrt{\lambda^2 - k^2}$ obtained by choosing the branch cut of Fig. 7.14 with $k = 1 + 0.5j$.

7.2.2.2 Im(μ) > 0

The curves of discontinuity are obtained by setting Re(μ) = 0. That is the case if

$$2uv - |k|^2 \sin 2\beta = 0 \text{ and } u^2 - v^2 - k^2 \cos 2\beta > 0. \quad (7.40)$$

After some calculations we obtain that the discontinuity curve is given by

$$|u| > |k| \cos\beta,\, v = \frac{|k|^2 \sin 2\beta}{2u}. \quad (7.41)$$

Similarly to the previous case, for $0 < \beta < \pi/2$, the curve will occupy the first and third quadrants, while for $-\pi/2 < \beta < 0$, the second and fourth quadrants. The curve has an asymptotic behavior toward infinity

$$v \to 0^+,\, \lambda \to \text{sgn}(\sin\beta)\infty \text{ and } v \to 0^-,\, \lambda \to -\text{sgn}(\sin\beta)\infty. \quad (7.42)$$

Figure 7.16 shows the proper branch cuts for $0 < \beta < \pi/2$ and illustrates the angles ϕ_1 and ϕ_2 as well as the regions where Re(μ) > 0 and Re(μ) < 0. The principal branch is the one with $n = 0$. For the case $-\pi/2 < \beta < 0$, the branch cuts are images of the ones from Fig. 7.16 respect to the u-axis.

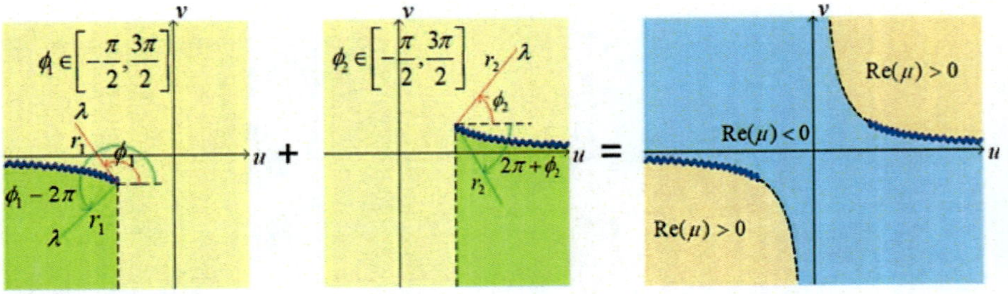

Fig. 7.16 Illustration of the branch cuts for $\mathrm{Im}(\mu) = 0$ and k complex

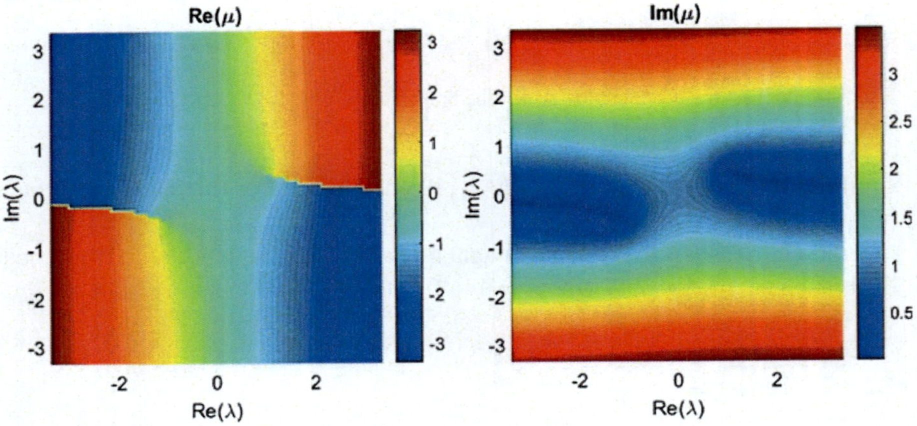

Fig. 7.17 Values of $\mathrm{Re}(\mu)$ and $\mathrm{Im}(\mu)$ for the branch cut of Fig. 7.16

Figure 7.17 shows the real and imaginary parts of $\sqrt{\lambda^2 - k^2}$ obtained by choosing the branch cut of Fig. 7.16 with $k = 1 + 0.5j$.

Chapter 8
Realization of Complex Sources

The most interesting feature of complex sources in comparison with real sources is the focused radiation that they can generate. Hence, in this chapter the realization of those type of sources is investigated.

8.1 Sound Focusing—Beamwidth

Complex monopoles have a principal lobe of radiation, while other high order multipoles have a principal lobe and side lobes as will be seen later. With increasing imaginary part, the main lobe of the complex monopole becomes narrower, hence the complex monopole alone can be used to produce a focused radiation. On the contrary, a focused radiation using real monopoles is possible only with a large number of sources and appropriate phase differences. A measure of the radiation focusing is the angular beamwidth D_B, defined as the angle θ at which the pressure level on the surface of a sphere centered in the source has reduced in 3 dB with respect to the level at the lobe axis, $\theta = 0°$. The angle θ is defined as

$$\cos\theta = \frac{\hat{u}_c \cdot \vec{R}}{R} \tag{8.1}$$

with $\vec{R} = \vec{r} - \vec{r}_q$, $R = \sqrt{\vec{R} \cdot \vec{R}}$.

M. Ochmann and R. Piscoya, *Theory and Application of Acoustic Sources Using Complex Analysis*, https://doi.org/10.1007/978-981-33-6040-2_8

The sound pressure is given by

$$p = \frac{\mathrm{e}^{-jkR_c}}{4\pi R_c}, \tag{8.2}$$

where $\vec{R}_c = \vec{r} - \vec{r}_q^{\,c}$ and $R_c = \sqrt{\vec{R}_c \cdot \vec{R}_c} = R_r + jR_i$.

The real and imaginary parts of $\vec{R}_c$ can be expressed explicitly in terms of θ

$$R_r(\theta) = \sqrt{\frac{\sqrt{R^4 + \beta^4 + 2\beta^2 R^2 \cos\theta} + R^2 - \beta^2}{2}}$$

$$R_i(\theta) = \sqrt{\frac{\sqrt{R^4 + \beta^4 + 2\beta^2 R^2 \cos\theta} - R^2 + \beta^2}{2}}$$

and the absolute square of R_c is simply

$$|R_c(\theta)|^2 = \sqrt{R^4 + \beta^4 + 2\beta^2 R^2 \cos 2\theta} \tag{8.3}$$

Then, the square of the pressure can be rewritten as

$$|p(\theta)|^2 = \frac{\mathrm{e}^{2kR_i(\theta)}}{16\pi^2} \frac{1}{\sqrt{R^4 + \beta^4 + 2\beta^2 R^2 \cos 2\theta}} \tag{8.4}$$

The ratio of the pressure squared at an arbitrary θ and at the axis of a complex monopole I_N is given by

$$I_N = \frac{|p(\theta)|^2}{|p(0)|^2} = \mathrm{e}^{-2k(R_i(0) - R_i(\theta))} \frac{R^2 + \beta^2}{\sqrt{R^4 + \beta^4 + 2\beta^2 R^2 \cos 2\theta}}, \tag{8.5}$$

A reduction of 3 dB in the pressure level is equivalent to a halving of the amplitude. It is not possible to find explicitly the angle θ at which $I_N = 0.5$, so it has to be obtained numerically. To do so, the root of the non-linear function

$$F(\theta) = 2\mathrm{e}^{-2k(R_i(0) - R_i(\theta))} - \frac{\sqrt{R^4 + \beta^4 + 2\beta^2 R^2 \cos 2\theta}}{R^2 + \beta^2}, \tag{8.6}$$

is sought.

Figure 8.1 shows the variation of the angular beamwidth of a complex monopole with the distance for different values of β and at two different frequencies. The beamwidth is constant beyond a distance larger than β.

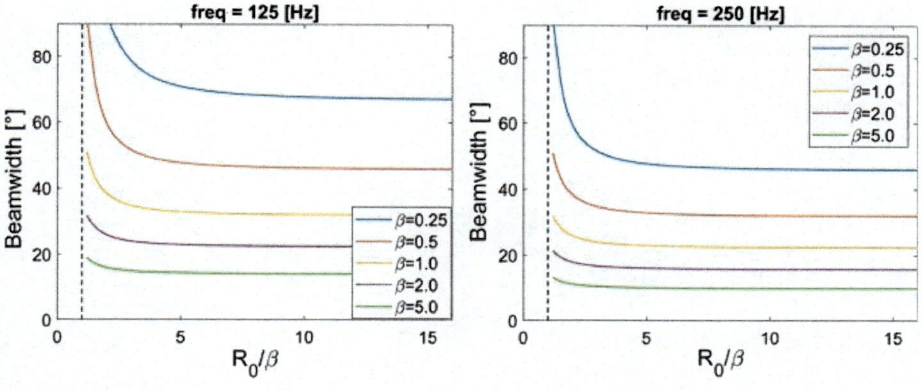

Fig. 8.1 Angular beamwidth of a complex monopole at two low frequencies

8.2 Integral Representation of a Monopole

According to Norris [Nor86], the sound pressure due to a monopole placed at $\vec{y}$ for a time-dependent factor $\mathrm{e}^{j\omega t}$ can be expressed in terms of the surface integral over a sphere of radius a and centered at $\vec{y}$.

$$\frac{\mathrm{e}^{-jk|\vec{x}-\vec{y}|}}{4\pi|\vec{x}-\vec{y}|} = \frac{1}{4\pi\, j_0(ka)} \int_0^{2\pi} \mathrm{d}\phi \int_0^{\pi} \frac{\mathrm{e}^{-jk|\vec{x}-(\vec{y}+\vec{a}(\theta,\phi))|}}{4\pi|\vec{x}-(\vec{y}+\vec{a}(\theta,\phi))|} \sin\theta \mathrm{d}\theta, \quad r > a \quad (8.7)$$

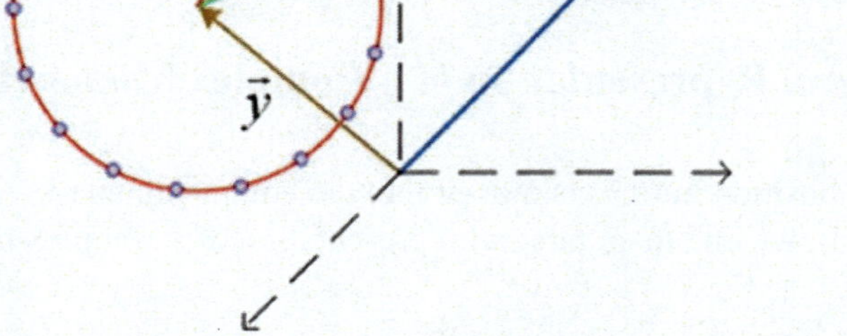

Fig. 8.2 Distribution of equivalent monopoles for a monopole representation

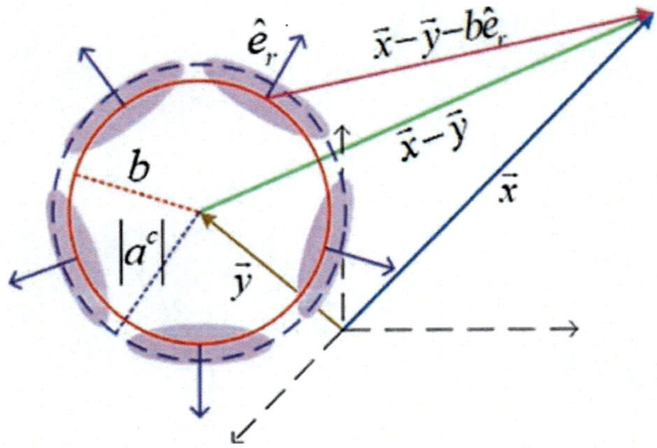

Fig. 8.3 Distribution of equivalent complex monopoles for a monopole representation

where $\vec{a}(\theta, \phi) = a\hat{e}_r(\theta, \phi)$ and $\hat{e}_r(\theta, \phi)$ is the radial vector in spherical coordinates of a coordinate system with origin at the center of the sphere (Fig. 8.2). Formula (8.7) is valid for field points outside the sphere.

According to formula (8.7), the sound pressure of a real monopole can be seen as the sound field due to a distribution of monopoles placed on the surface of the sphere with radius a and uniform amplitude density $\left[4\pi a^2 j_0(ka)\right]^{-1}$.

By letting a be complex, $a^c = b - j\beta$, formula (8.7) provides an integral representation of a monopole in terms of complex monopoles. In that case, this representation is valid for $r > |a^c|$.

$$\frac{\mathrm{e}^{-jk|\vec{x}-\vec{y}|}}{4\pi|\vec{x}-\vec{y}|} = \frac{1}{4\pi\, j_0(ka^c)} \int_0^{2\pi} \mathrm{d}\phi \int_0^{\pi} \frac{\mathrm{e}^{-jk|\vec{x}-\vec{a}^c(\theta,\phi)|}}{4\pi|\vec{x}-\vec{a}^c(\theta,\phi)|} \sin\theta \mathrm{d}\theta, \quad r > \left|a^c\right| \tag{8.8}$$

Therefore, the monopole can be represented by a distribution of complex monopoles centered at the complex positions $\vec{a}^c = (b - j\beta)\hat{e}_r$, and uniform amplitude density $\left[4\pi a^2 j_0(ka^c)\right]^{-1}$. The equivalent complex monopoles have varying unit vector $\hat{u} = \hat{e}_r$. A graphic representation of the distribution is given in Fig. 8.3.

8.3 Integral Representation of a Complex Monopole

If the source position in (8.7) is moved into the complex plane $\vec{y} \to \vec{y}^c = \vec{y} - j\beta\,\hat{u}$ (a is kept real), we obtain an integral representation of a complex monopole.

$$\frac{\mathrm{e}^{-jk|\vec{x}-\vec{y}^c|}}{4\pi|\vec{x}-\vec{y}^c|} = \frac{1}{4\pi\, j_0(ka)} \int_0^{2\pi} \mathrm{d}\phi \int_0^{\pi} \frac{\mathrm{e}^{-jk|\vec{x}-\vec{s}^c(\theta,\phi)|}}{4\pi|\vec{x}-\vec{s}^c(\theta,\phi)|} \sin\theta \mathrm{d}\theta \tag{8.9}$$

The sound field of a complex monopole at $\vec{y}^c$ can be seen as the field due to a collection of complex monopoles at varying complex positions $\vec{s}^c(\theta, \phi) = \vec{y}^c + \vec{a}(\theta, \phi)$.

In this case, the range where the integral is valid depends on the direction of the field points and $\hat{u}$ but a safe value is $r > a + \beta$. A graphic representation of the distribution is given in Fig. 8.4.

Expression (8.9) was numerically computed for a complex monopole with unit vector $\hat{u} = (1, 0, 0)$. Three different sets of field points were considered. The first set contains points at the $\hat{u}$-axis, the second one contains points at a line perpendicular to $\hat{u}$ and the third one contains points in a plane parallel to $\hat{u}$ (Fig. 8.5). Nine different values of a were evaluated.

Figure 8.6 shows a comparison between the integral representation of the complex monopole and the exact expression at field points on the $\hat{u}$-axis for different values of a. A perfect agreement is obtained for $r > a$.

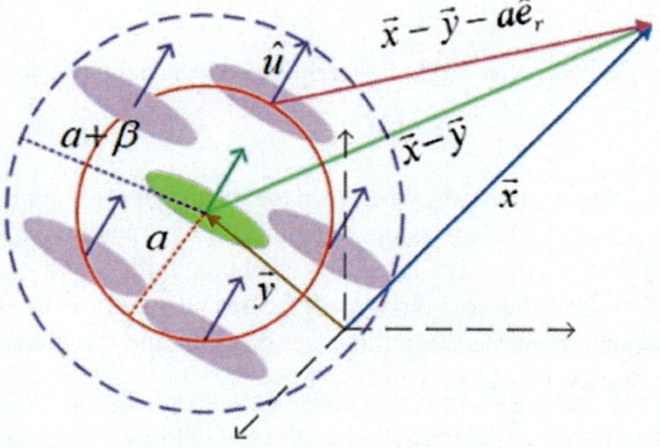

Fig. 8.4 Distribution of equivalent complex monopoles for a complex monopole representation

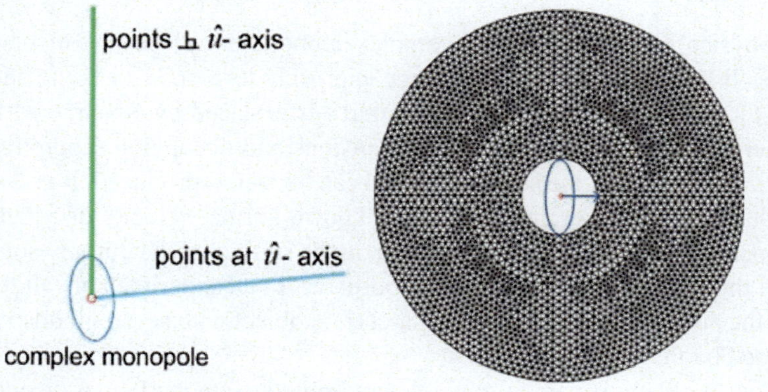

Fig. 8.5 Field points on two lines (left) and on a plane (right)

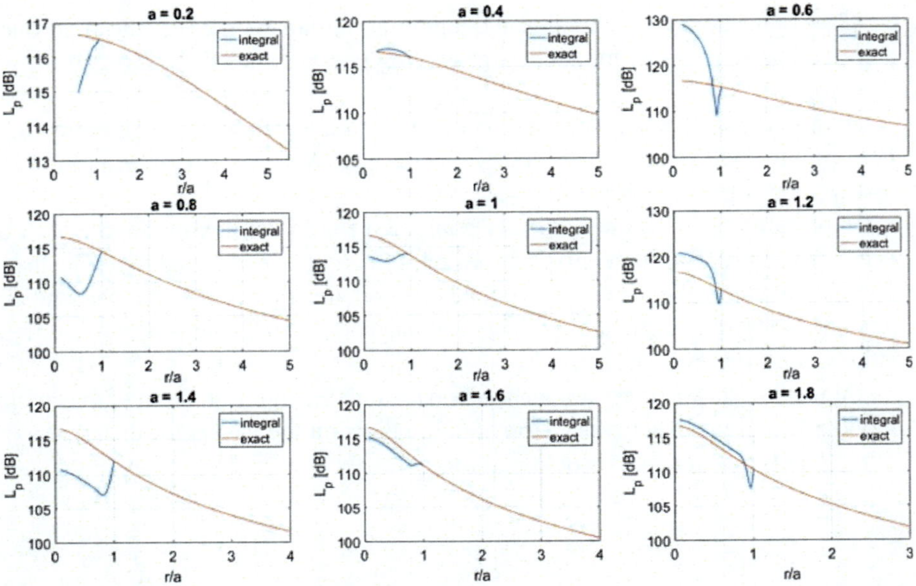

Fig. 8.6 Comparison between exact and integral representation of complex monopole for field points on the $\hat{u}$-axis

Figure 8.7 shows a comparison between the integral representation of the complex monopole and the exact expression at field points perpendicular to the $\hat{u}$-axis for different values of a. A perfect agreement is obtained for $r > a + \beta$.

Figure 8.8 shows the comparison of the sound pressure in the plane XY. As expected, the agreement between the exact solution and the integral representation is very good outside the radius $r = a + \beta$.

8.4 Physical Realization of a Complex Monopole

The physical realization of the complex monopole will be investigated using an approach similar to the dummy source approach developed by Pavić and Lindberg [Pav16]. In that method, the original source is replaced by a source with a simpler geometry (e.g., a box) but similar dimensions and occupying (approximately) the same volume, so that the scattered sound can be well reproduced. It is assumed, that a certain number of patches (D), with a known normal velocity distribution (e.g., a loudspeaker membrane), are distributed on the surface of the dummy source and the rest of the surface is rigid. This assumption makes an experimental validation easier, since the dummy source can be built as a rigid object with several loudspeakers flush mounted on the walls.

In our problem, we are in principle not limited to any particular object size, since our original source is a mathematical construction. The only possible limitation is

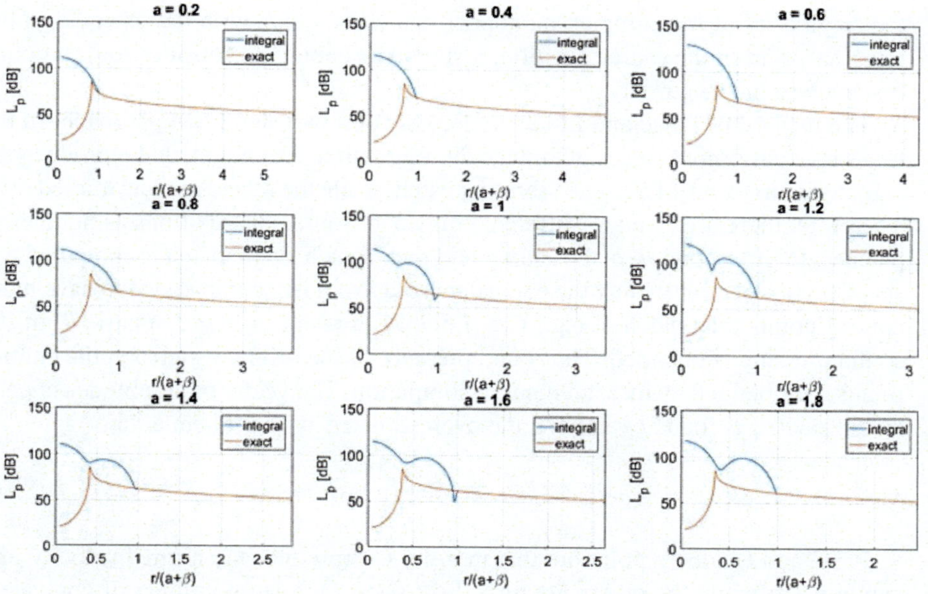

Fig. 8.7 Comparison between exact and integral representation of complex monopole for field points perpendicular to the $\hat{u}$-axis

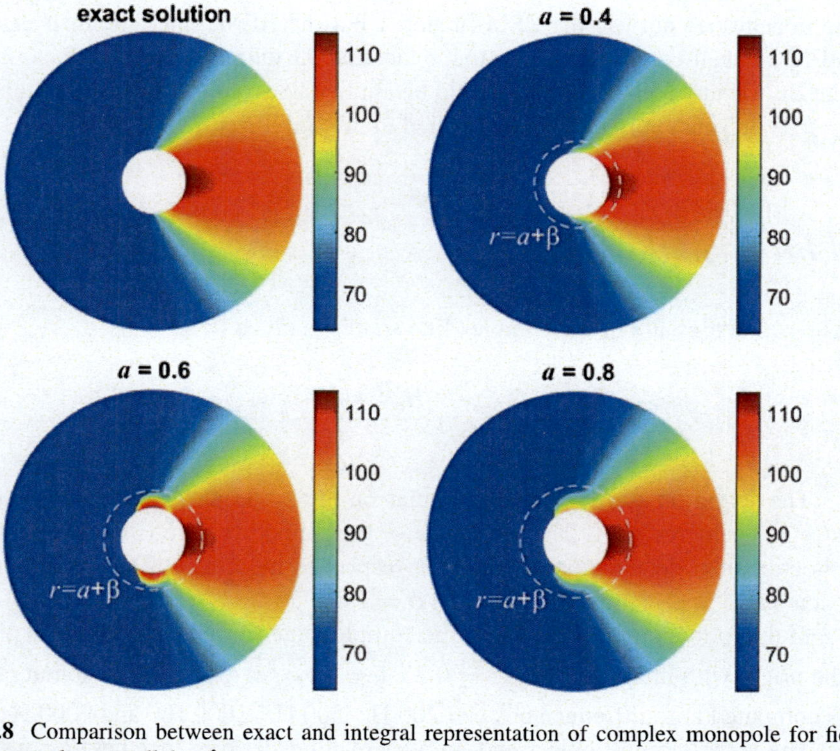

Fig. 8.8 Comparison between exact and integral representation of complex monopole for field points on a plane parallel to $\hat{u}$

the singularity of the complex monopole, which is a disk with radius equal to the imaginary part of the source position. It should be enough to consider control points that enclose the singularity.

The method of Pavić and Lindberg consists of a two-step ESM approach. In the first step, equivalent sources within the dummy source are sought, that reproduce the velocity distribution of a single vibrating patch, while the others remain "turned off." Therefore, there are as many different equivalent source distributions as number of patches. In the second step, the sound field due to each single patch is computed at a set of C control points using the equivalent sources. It is recommended to have more control points than patches, e.g., $C > 1.5D$. As a result, a $C \times D$ matrix $\boldsymbol{T}$ of the sound pressure is obtained. The sound pressure of the original source at the control points $\boldsymbol{p}$ is obtained by measurement or simulation. The vector of amplitudes of each single patch $\boldsymbol{q}$ is found by solving the overestimated system of equations

$$Tq = p \tag{8.10}$$

Pavić and Lindberg point out that in order to reproduce the normal velocity of a single patch while the others are turned off (step 1), a large number of equivalent sources is required. This leads to highly ill-conditioned matrices. In the second step, the number of patches is relatively small, which leads to a condition number of T not very large. To avoid the problems that the ESM causes in step 1, the present approach does not use the ESM in step 1 but the BEM, since the matrices of the BEM are well-conditioned. Due to the fact that the dummy source surface is closed, the Burton and Miller method should be applied to avoid the critical frequencies.

The present approach will be applied to 2D and 3D cases.

8.4.1 The 2D-Case

The sound pressure of the complex line source is given by

$$p(\vec{x}) = \frac{H_0^{(2)}(k|\vec{x} - \vec{y}^c|)}{4j} \tag{8.11}$$

The sound pressure becomes singular on the a line of length 2β perpendicular to $\hat{u}$ and centered on $\vec{y}$. The amplitude of the sound pressure is not the same in all directions but depends on the angle ϕ respect to the $\hat{u}$-axis. The maximum value occurs at $\phi = 0°$ and the minimum at $\phi = 180°$.

In the following numerical test, the complex line source investigated is placed at the origin with imaginary part $\beta = 0.25$, $\hat{u} = \left(1/\sqrt{2},\ 1/\sqrt{2}\right)$. The sound radiation is computed at four frequencies $f = 200$ Hz, 500 Hz, 1000 Hz, and 2000 Hz.

The same dummy object and number of loudspeakers (patches) as in [Pav16] are used. The box has 0.7 m length and 0.5 m width. The number of loudspeakers

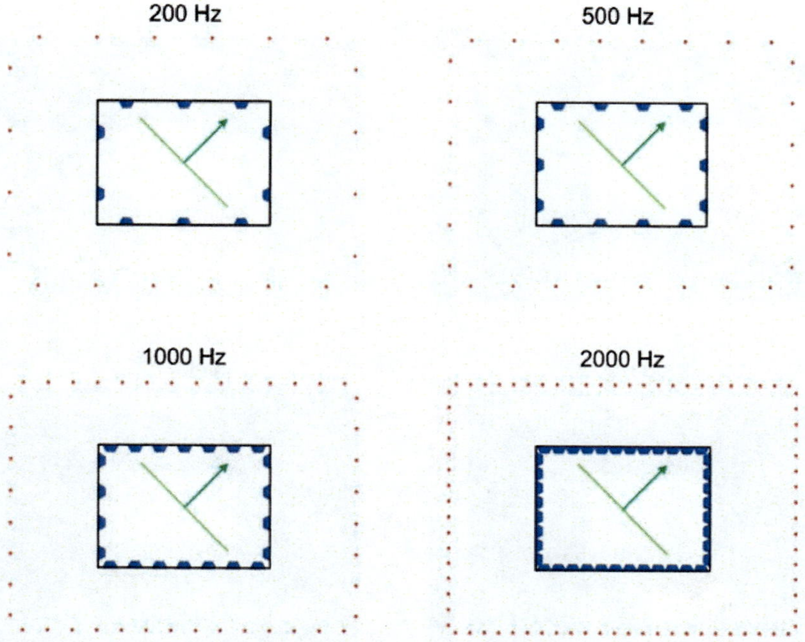

Fig. 8.9 Position of loudspeakers and control points for different frequencies and the complex line source with its singularity line

(patches) and control points depend on the frequency. The number of loudspeakers is 10, 14, 22, and 40 and the number of control points is 20, 28, 44, and 80 ($D = 2C$), respectively. The number of patches per wavelength varies between 10.3 at 200 Hz and 2.7 at 2000 Hz. The velocity distribution follows a cosine function.

The control points are placed on the surface of an imaginary box twice the size of the dummy source (Fig. 8.9).

With the BEM, the sound field due to each single patch is computed. Figure 8.10 shows the radiation of a patch at each frequency.

Once the pressure on the surface is known, the integral equation is applied to compute the sound pressure at the C microphone positions. Doing that for each patch, the $C \times D$ matrix $\boldsymbol{T}$ is built. The complex amplitude of the patches is calculated solving the overestimated system of equations. Figure 8.11 shows a comparison between the sound pressure of the dummy source (left) and the exact complex line source (right) at points outside the dummy source.

Figure 8.11 shows that the error is larger near the surface of the dummy source but after some small distance, the agreement is very good. To see if that holds in the far field, the radiation pattern on a circle of radius 10 m has been calculated. Figure 8.12 shows a comparison of the radiation patterns between dummy source and analytical solution.

The polar plots show a variation of −40 dB respect to the maximum value. A greater difference is not perceivable. For that range, the agreement is excellent. It is important to mention that the variation in pressure level of the analytical solution

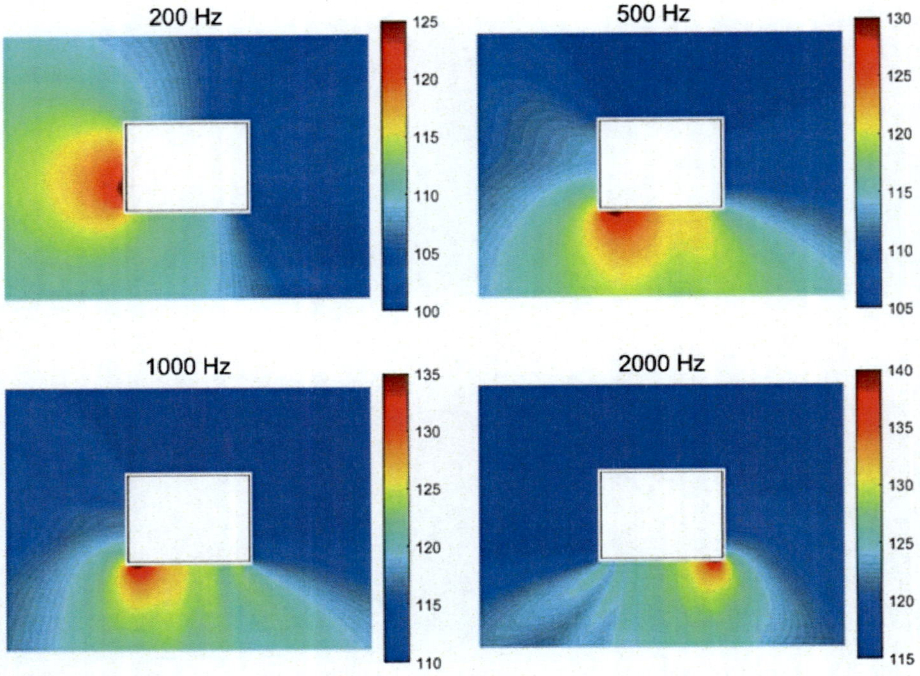

Fig. 8.10 Radiation of a single patch on the dummy for different frequencies

can reach -80 dB at 1000 Hz and -160 dB at 2000 Hz. Such differences cannot be reached by numerical simulations.

8.4.2 The 3D-Case

The application of the dummy source method in 3D follows the same strategy as in 2D. The inclusion of a third dimension leads to an increase in the size of the mesh and the number of surface patches (loudspeakers). The three-dimensional Green's function for the BEM approach has to be considered. The sound pressure of the complex monopole is given by

$$p(\vec{x}) = \frac{\mathrm{e}^{-jk|\vec{x}-\vec{y}^c|}}{4\pi\,|\vec{x}-\vec{y}^c|} \tag{8.12}$$

We are looking for a dummy source that can reproduce the radiation pattern of a complex monopole in any direction. Therefore, a spherical like radiator was considered. Three different models were investigated: (a) dodecahedron (12 faces), (b) icosahedron (20 faces), and (c) truncated icosahedron (soccer ball, 32 faces). The three models have an approximated diameter of 50 cm. In each face there is a

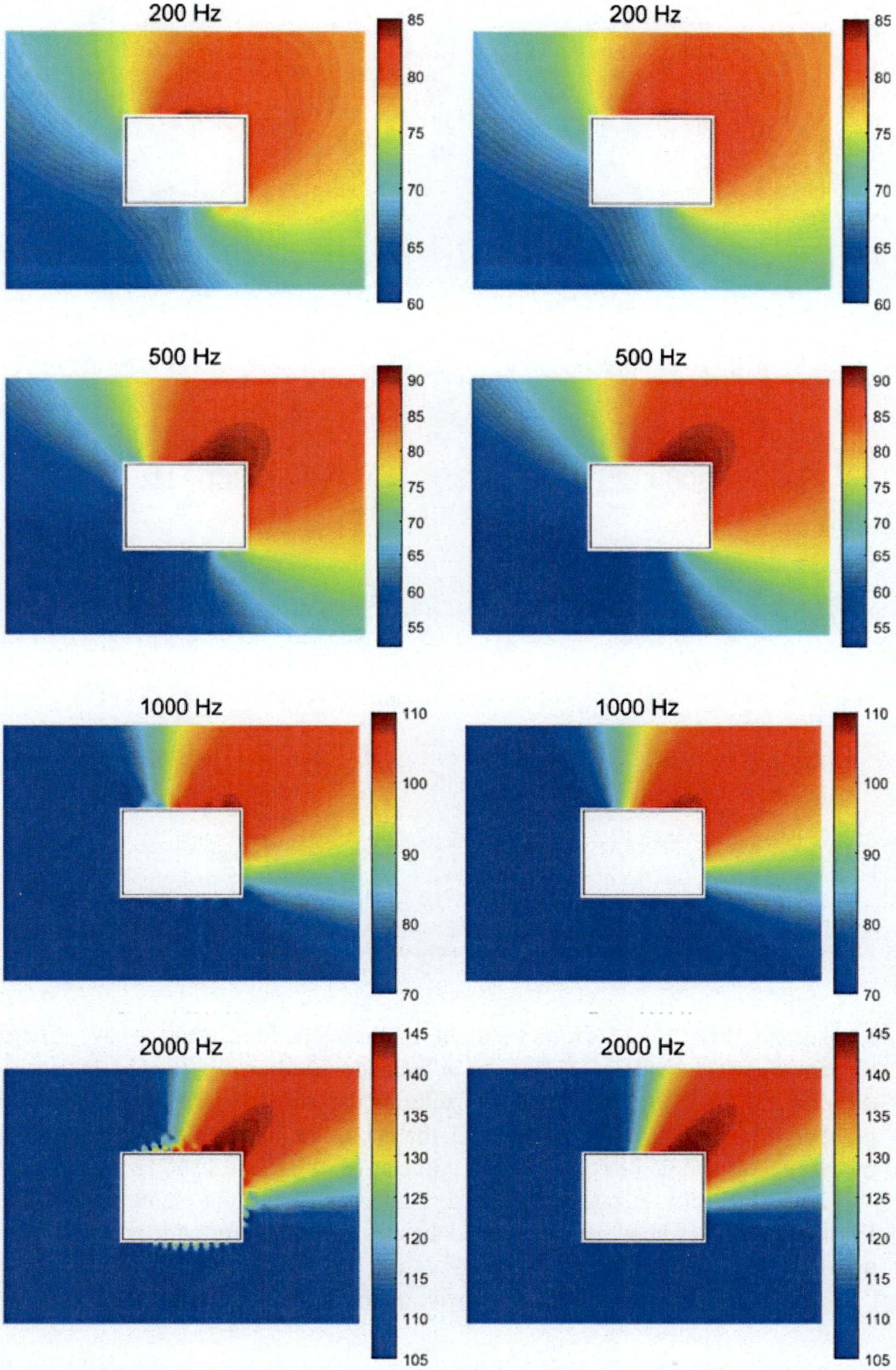

Fig. 8.11 Radiated pressure of the dummy source (left) and the complex line source (right)

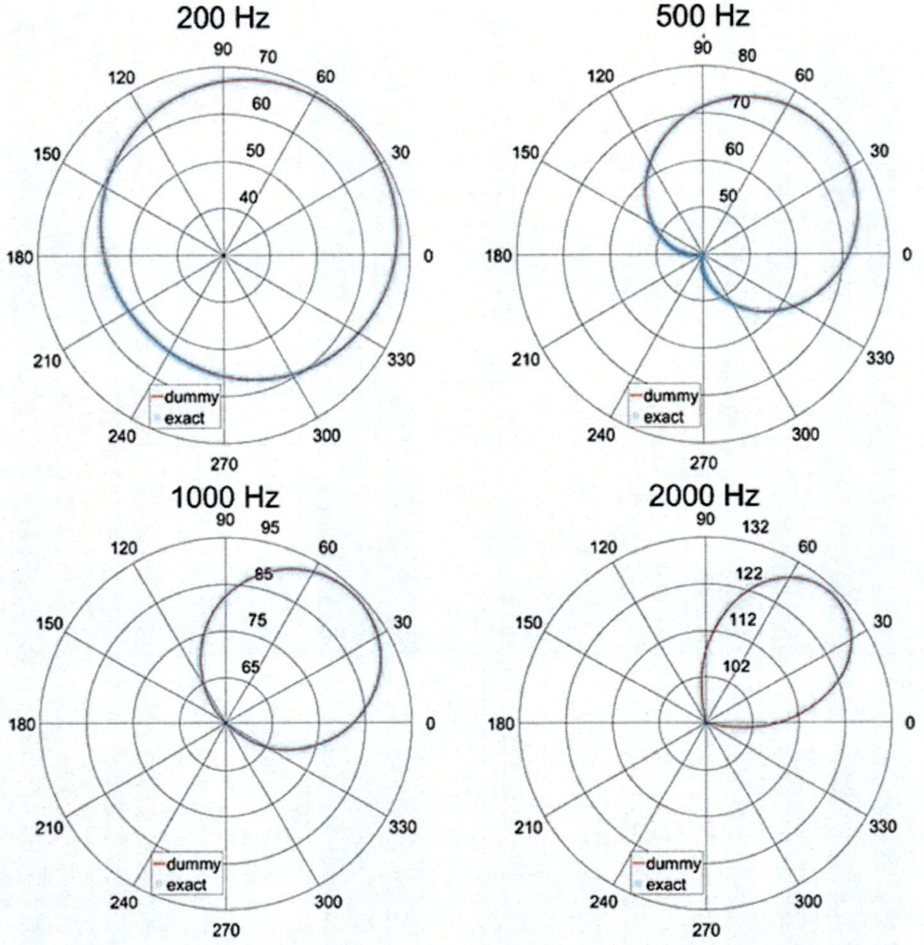

Fig. 8.12 Radiation pattern of the complex line source

loudspeaker. Figure 8.13 shows the three models. The loudspeakers have different diameters in each model, D_L: 18.3 cm in the dodecahedron, 13.7 cm in the icosahedron, and 11 cm in the soccer ball. Considering the suggestion from Pavic about the number of loudspeakers per wavelength, the maximum frequency that can be solved by the models should be $f_{\text{rep}} = c/2D_L$: 930 Hz, 1250 Hz, and 1560 Hz, respectively.

The surface discretization was performed considering ten elements per wavelength. Figure 8.14 shows the generated meshes that should provide accurate results up to 2000 Hz.

The loudspeakers have a radial velocity distribution of the form

$$v_n = \frac{n+1}{\pi (D_L/2)^2}\left[1 - \left(\frac{r}{D_L/2}\right)^2\right]^n \tag{8.13}$$

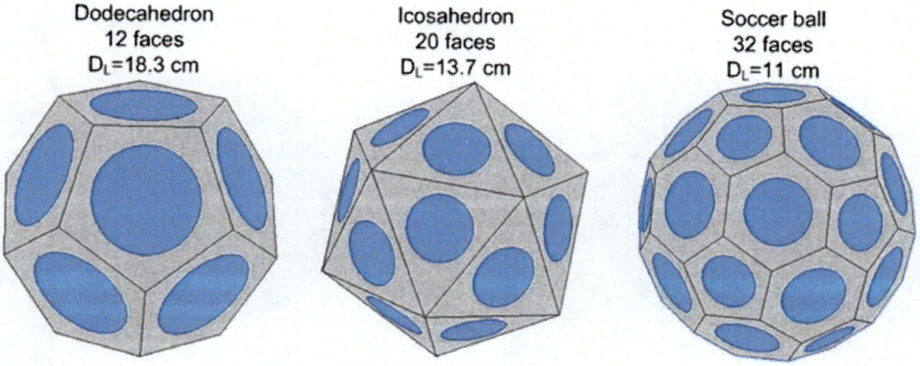

Fig. 8.13 Different models for the 3D dummy source

Fig. 8.14 Mesh of the dummy source models

which ensures a unit volume velocity. In our examples $n = 2$ (see Fig. 8.15).

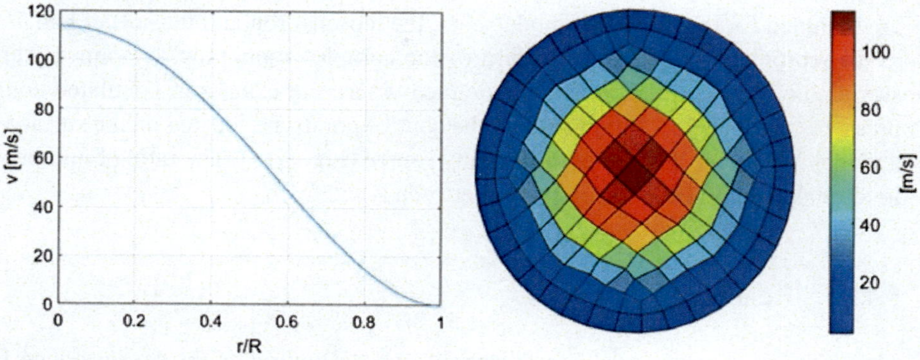

Fig. 8.15 Velocity distribution of the loudspeakers

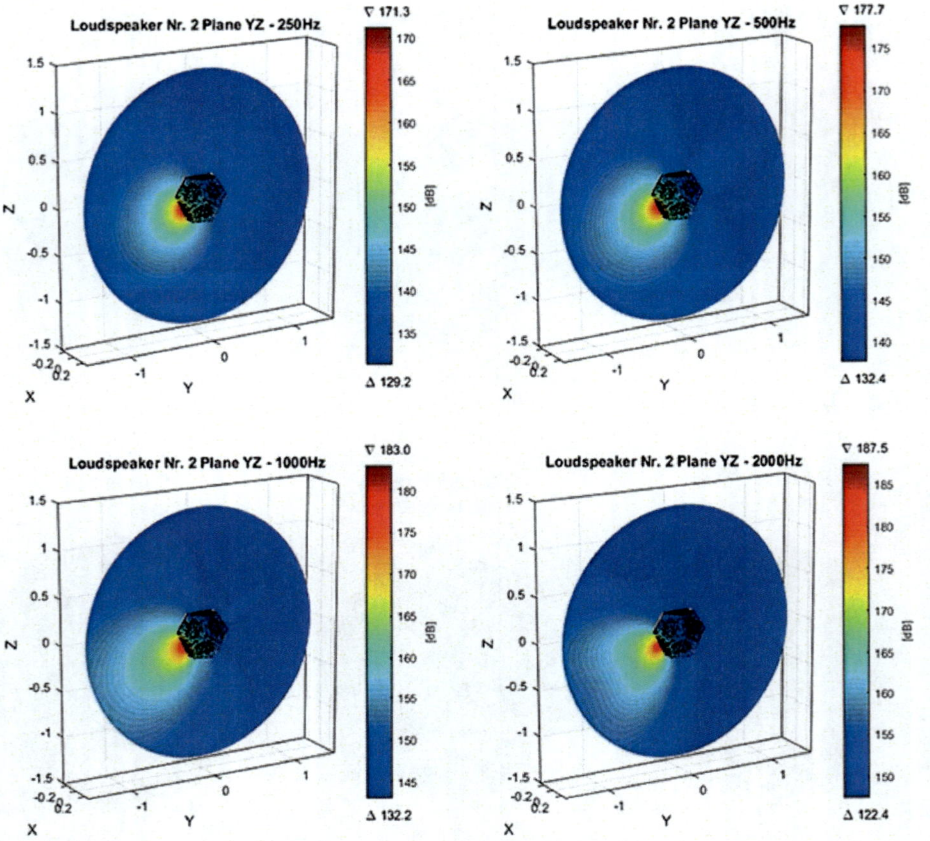

Fig. 8.16 Sound radiation of one loudspeaker in the dodecahedron

Figure 8.16 shows the sound radiation of one loudspeaker turned on while the others are turned off in the dodecahedron model at four frequencies. Similar results are obtained for the other two models, i.e., the icosahedron and the soccer ball.

To reproduce the radiation pattern of the complex monopole, C control points surrounding the dummy source are defined where the exact and simulated sound pressures are matched. For our study, the control points are located on the surface of a model with twice the size of the dummy source (Fig. 8.17). The ratio of number of loudspeakers D to control points lies between 4 and 6.

8.4.2.1 Results

enlargethispage-1pcTo evaluate the quality of the reproduction we introduce the mean relative error at the control positions ε_C:

Fig. 8.17 Control points for each dummy source model

$$\varepsilon_C = \sqrt{\frac{\sum_{i=1}^{C} \left| p_i^{(s)} - p_i^{(e)} \right|^2}{\sum_{i=1}^{C} \left| p_i^{(e)} \right|^2}} \times 100 \tag{8.14}$$

where $p_i^{(s)}$ and $p_i^{(e)}$ are the simulated and exact pressures at the i-th control point.

We have performed a simulation for a complex monopole with imaginary amplitude $\beta = 0.2$ with the three models. This value of β ensures that the singularity ring of the complex monopole remains inside the dummy source. As mentioned before, the control points are placed on the boundary of a surface with the same form of the dummy source but with twice its size. Therefore, an agreement between simulation and exact values is expected in a region outside the control points.

Table 8.1 shows the errors ε_C obtained using the described models. The errors below 10% are written in bold and the errors above 10% in italic. As one may expect, the error decreases with increasing number of loudspeakers when the model has two or more loudspeakers per wavelength $f \leq 1000$ Hz.

Figures 8.18 and 8.19 show the sound pressure radiated by the dummy sources and the exact analytical sound pressure. At 500 Hz (Fig. 8.18), the three models show a good reproduction of the radiation of the complex monopole since the relative errors lie below 10%. At 1000 Hz (Fig. 8.19), only the soccer ball model with 32 loudspeakers is able to reproduce accurately the monopole sound radiation with an error $\varepsilon_C = 4.9$.

Table 8.1 Relative error at the control points ε_C

	125 Hz	250 Hz	500 Hz	1000 Hz	2000 Hz
12 Lspk	**2.6**	**3.3**	**9.3**	*54.2*	*66.2*
20 Lspk	**1.5**	**1.9**	**5.8**	*42.7*	*78.5*
32 Lspk	**0.3**	**0.3**	**0.5**	**4.9**	*72.7*

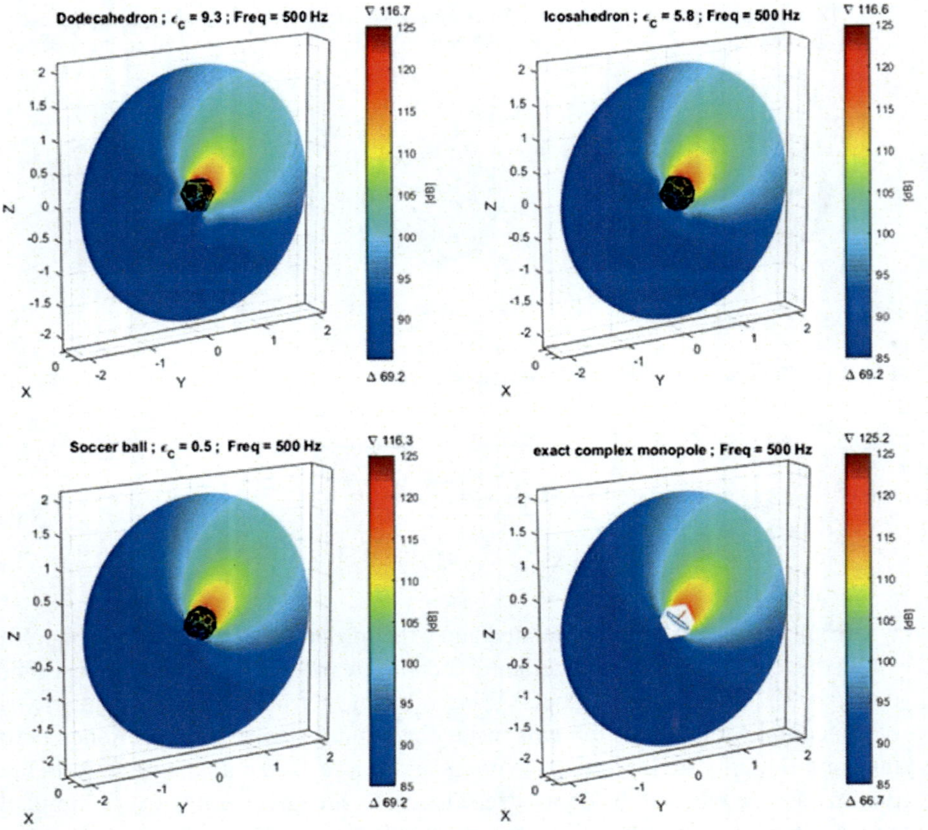

Fig. 8.18 Sound radiation from dummy sources and analytical solution at 500 Hz

8.4.2.2 Singularity Larger Than Dummy Source

If the singularity of the complex monopole must be always inside the dummy source, the size of the dummy will limit the amplitude of the imaginary part to be reproduced. We investigate how does the reproduction behaves, if the singularity is larger. Two values of β, $\beta = 0.5$ and $\beta = 1.0$ and the soccer ball model with 32 loudspeakers are considered. Table 8.2 shows the relative error obtained. The results at 2000 Hz are not taken into account since the model was not able to provide accurate results at this frequency by $\beta = 0.2$. In all cases, the relative error is too large.

We repeat the calculation but this time, the position of the control points is moved away with a factor A_f respect to their original position ($A_f = 1$) in order to place them always beyond the singularity. Tables 8.3 and 8.4 show the results for different A_f and both β. For $\beta = 0.5$, placing the control points beyond the singularity allows to obtain accurate results up to 500 Hz. The ratio of reduction of the error is slow with increasing A_f. For $\beta = 1.0$, accurate results are obtained only up to 250 Hz and the ratio of reduction of the error with increasing A_f is again very slow.

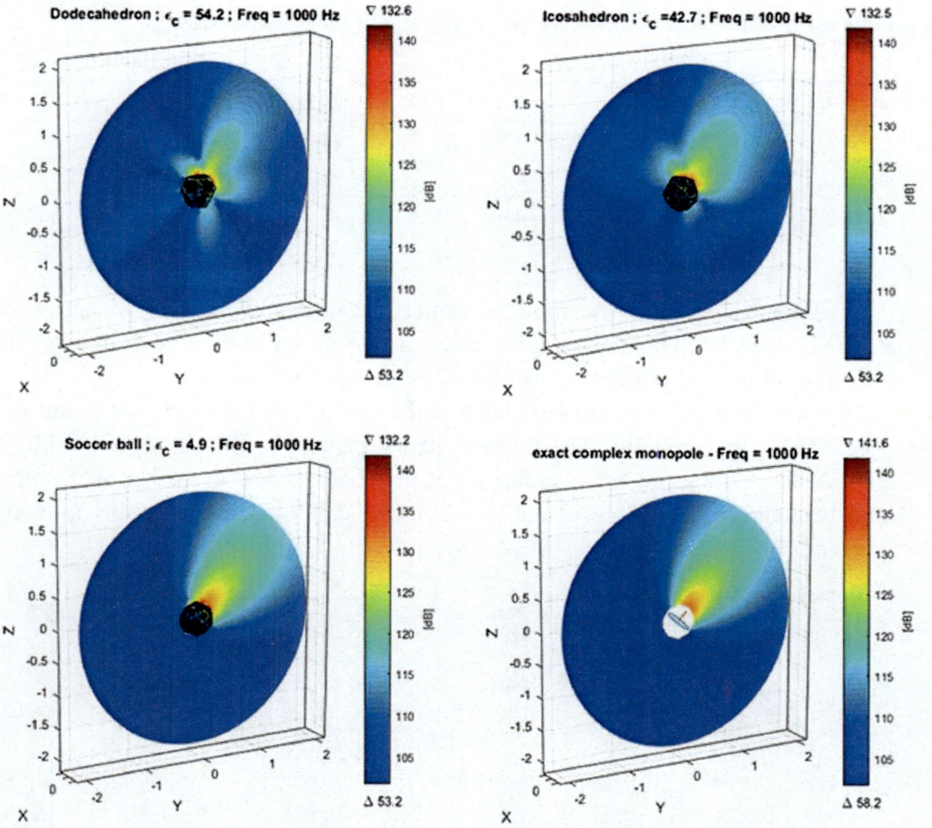

Fig. 8.19 Sound radiation from dummy sources and analytical solution at 1000 Hz

Table 8.2 Relative error at the control points ε_C when the singularity is larger than the model

	125 Hz	250 Hz	500 Hz	1000 Hz
$\beta = 0.5$	*44.5*	*27.1*	*12.5*	*22.9*
$\beta = 1.0$	*21.3*	*19.7*	*23.1*	*48.8*

Table 8.3 Relative error at the control points ε_C for $\beta = 0.5$ and different A_f

	125 Hz	250 Hz	500 Hz	1000 Hz
$A_f = 2$	**0.7**	**1.0**	**4.9**	*20.4*
$A_f = 3$	**0.1**	**0.6**	**4.6**	*20.1*
$A_f = 4$	**0.1**	**0.5**	**4.5**	*20.0*
$A_f = 5$	**0.0**	**0.5**	**4.4**	*20.0*

Table 8.4 Relative error at the control points ε_C for $\beta = 1.0$ and different A_f

	125 Hz	250 Hz	500 Hz	1000 Hz
$A_f = 3$	**2.7**	**5.6**	*19.0*	*42.3*
$A_f = 4$	**1.0**	**4.8**	*18.5*	*42.4*
$A_f = 6$	**0.6**	**4.5**	*18.2*	*42.5*
$A_f = 8$	**0.5**	**4.4**	*18.1*	*42.5*

Figure 8.20 compares the simulated and exact pressures for $\beta = 0.5$ and $A_f = 2$ at 500 Hz and 100 Hz. The relative errors are $\varepsilon_C = 4.9$ and $\varepsilon_C = 20.4$ (see Table 9.3). The white dots show the position of the control points.

Figure 8.21 compares the simulated and exact pressures for $\beta = 1.0$ and $A_f = 3$ at 250 Hz and 500 Hz. The relative errors are $\varepsilon_C = 5.6$ and $\varepsilon_C = 19.0$ (see Table 8.4). As a summary, we can point out that the soccer ball model with 32 loudspeakers is able to reproduce the radiation pattern of the complex monopole

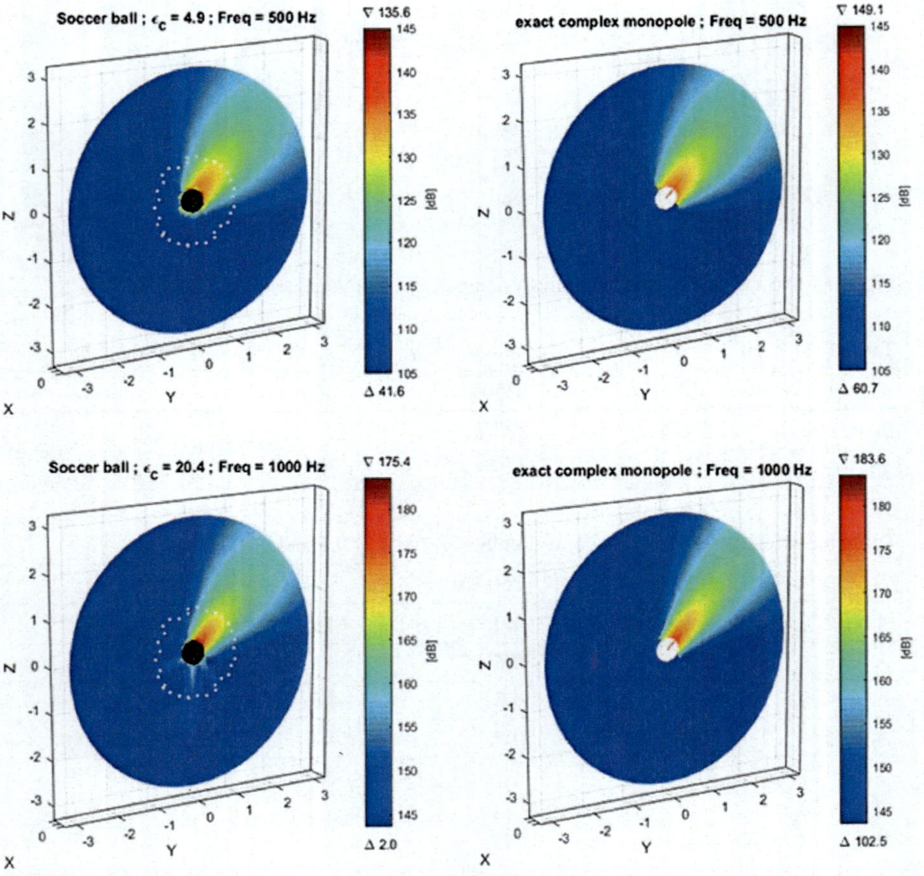

Fig. 8.20 Sound radiation from soccer ball model and analytical solution for $\beta = 0.5$ and $A_f = 2$

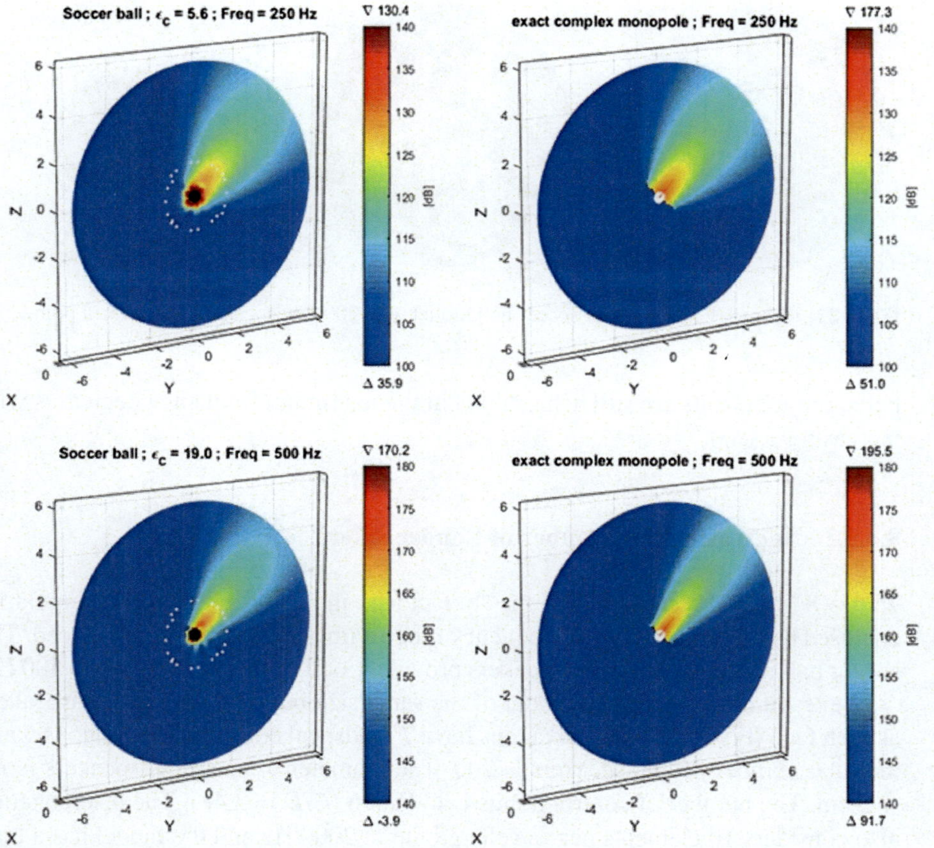

Fig. 8.21 Sound radiation from soccer ball model and analytical solution for $\beta = 1.0$ and $A_f = 3$

with a singularity completely enclosed by the model up to 1000 Hz. If the imaginary part is increased, the singularity becomes larger than the source model. It can still be reproduced accurately but the control points need to be moved away until they completely enclose the singularity. The increase in the imaginary part also moves the frequency range where the reproduction is possible to the low frequencies.

We summarize the results of the relative error obtained for the soccer ball model in Table 8.5. The number of loudspeakers should provide accurate results up to 1000 Hz, but that holds only when the singularity is enclosed in the model. For larger imaginary

Table 8.5 Relative error at the control points ε_C for different values of β and A_f for the soccer ball model with 32 loudspeakers

	125 Hz	250 Hz	500 Hz	1000 Hz
$\beta = 0.2, A_f = 1$	**0.3**	**0.3**	**0.5**	**4.9**
$\beta = 0.5, A_f = 2$	**0.7**	**1.0**	**4.9**	*20.4*
$\beta = 1.0, A_f = 3$	**2.7**	**5.6**	*19.0*	*42.3*

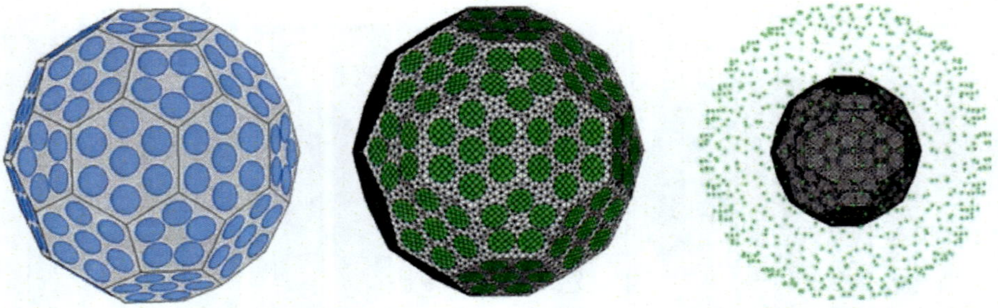

Fig. 8.22 Geometry and mesh of soccer ball model with 200 loudspeakers and control points

parts, proper results are still achievable but the maximum frequency decreases with increasing β (A_f).

8.4.2.3 Increasing the Number of Loudspeakers

Tables 8.3 and 8.4 show that if the A_f factor is increased to keep the singularity enclosed by control points, the frequency range with accurate results is reduced. The soccer ball model with 32 loudspeakers provides good results at $\beta = 1$ up to 500 Hz.

We test a new soccer ball model of the same size but with several loudspeakers at each face (Fig. 8.22). The hexagons have 7 loudspeakers and the pentagons 5 and the total number of loudspeakers is 200. The diameter of each loudspeakers is $D_L = 5$ cm. The number of control points is $C = 696$ ($C/D = 3.48$). The discretization also considers 10 elements per wavelength up to 2000 Hz and the model has a total of 10,966 elements. Considering the suggestion from Pavic about the number of loudspeakers per wavelength, the maximum frequency that can be solved by the models should be $f_{rep} = c/2D_L = 3430$ Hz.

The relative error ε_C is computed again for $\beta = 1$ and also for $\beta = 2$ and $\beta = 4$. The A_f factors are 3, 5, and 10, respectively. The values of ε_C are shown in Table 8.6. By the last two values of β, the matrix of the transfer functions for the considered set of control points has a deficient rank at 125 and 250 Hz but still useful results can be obtained.

Figures 8.23 and 8.24 show the sound radiation of the dummy source model for the complex monopoles with imaginary parts $\beta = 2$ and $\beta = 4$. From the results of Tables 8.5 and 8.6, one can see that in our case, the minimum required number of loudspeakers to obtain accurate results not only depend on the frequency but also on the magnitude of the imaginary part of the complex monopole to be reproduced.

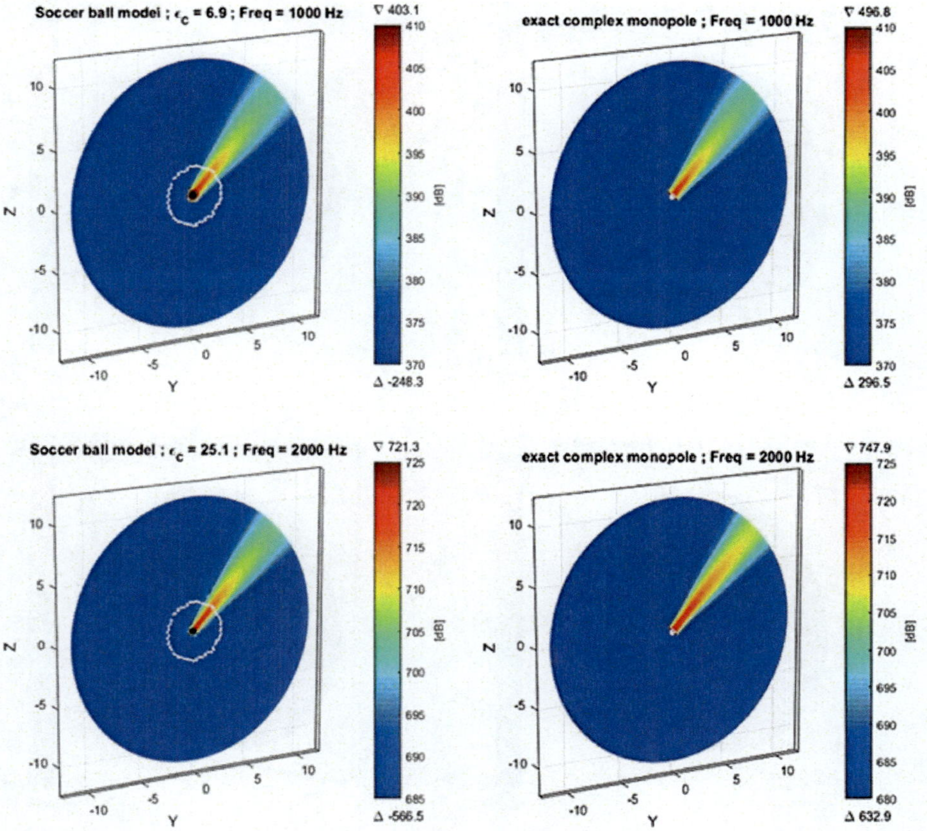

Fig. 8.23 Sound radiation from soccer ball model and analytical solution for $\beta = 2$ and $A_f = 5$

8.5 Complex Multipoles

As in the case of real multipoles, complex multipoles can be obtained from the complex monopole. Following [Shi77], we define the complex multipole of order (m, n, l) by differentiating the complex monopole repeatedly with respect to x, y, and z:

$$p_{nml}(\vec{r}, \vec{r}') = \frac{\partial^n}{\partial x^n} \frac{\partial^m}{\partial y^m} \frac{\partial^l}{\partial z^l} p(\vec{r}, \vec{r}') \tag{8.15}$$

Such representation is easy to obtain for the first orders but becomes unpractical for higher orders. An alternative and more practical representation is the use of spherical basis functions with complex source positions:

$$\psi^c_{nm} = h_n^{(2)}(k|\vec{r} - \vec{r}'|) P_n^m(\cos\gamma) \cos m\delta \tag{8.16a}$$

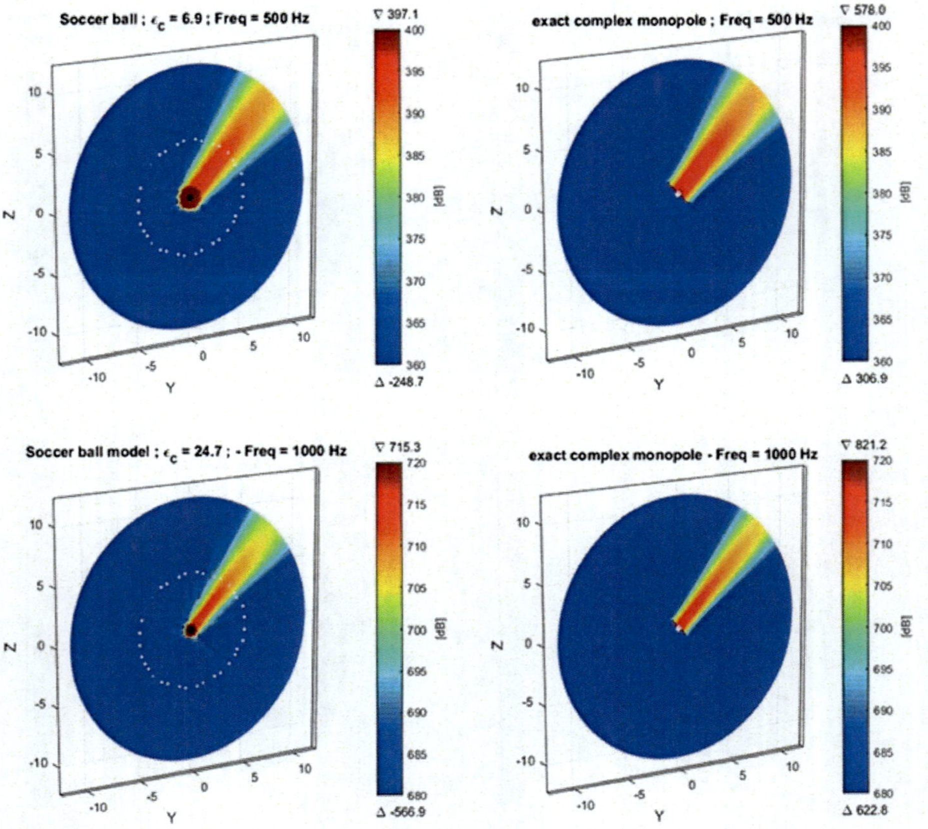

Fig. 8.24 Sound radiation from soccer ball model and analytical solution for $\beta = 4, A_f = 10$

Table 8.6 Relative error at the control points ε_C for $\beta = 1$, 2, and 4 with the soccer ball model with 200 loudspeakers

	125 Hz	250 Hz	500 Hz	1000 Hz	2000 Hz
$\beta = 1, A_f = 3$	**0.0**	**0.0**	**0.0**	**0.7**	**7.2**
$\beta = 2, A_f = 5$	**0.2***	**0.1***	**0.7**	**6.9**	*25.1*
$\beta = 4, A_f = 10$	**0.3***	**1.1***	**6.9**	*24.7*	*50.0*

*Matrix with deficient rank

$$\psi^s_{nm} = h_n^{(2)}(k|\vec{r} - \vec{r}'|)P_n^m(\cos\gamma)\sin m\delta \tag{8.16b}$$

where $\cos\gamma = \frac{z-z'}{|\vec{r}-\vec{r}'|}$ and $\tan\delta = \frac{y-y'}{x-x'}$. For the first orders of the multipole, the relation between the expressions in (8.15) and (8.16) can be deduced:

Monopole:

$$\psi_{00}^{c} = \frac{j4\pi}{k} p_{000}. \tag{8.17}$$

Dipoles

$$\begin{aligned} \psi_{10}^{c} &= -\frac{j4\pi}{k^2} p_{001}. \\ \psi_{11}^{c} &= \frac{j4\pi}{k^2} p_{100},\ \psi_{11}^{s} = \frac{j4\pi}{k^2} p_{010}, \end{aligned} \tag{8.18}$$

Quadrupoles

$$\begin{aligned} \psi_{20}^{c} &= \frac{j6\pi}{k^3} p_{002} + \frac{j2\pi}{k} p_{000} \\ \psi_{21}^{c} &= -\frac{j12\pi}{k^3} p_{101},\ \psi_{21}^{s} = -\frac{j12\pi}{k^3} p_{011}, \\ \psi_{22}^{c} &= \frac{j12\pi}{k^3} (p_{200} - p_{020}),\ \psi_{22}^{s} = \frac{j24\pi}{k^3} p_{110}. \end{aligned} \tag{8.19}$$

There is also an approximation for the region near the positive z-axis in terms of Hermite-Gaussian beam modes (see [Shi77]). For $\vec{\beta} = (0, 0, c)$, the multipole approximated function is given by:

$$G_{nm}(\vec{r}) \approx \frac{2}{k} e^{kc} \frac{(-1)^{m+n}}{4\pi} e^{j(m+n)\pi/4} e^{-j(kz+\tau^2(x^2+y^2))} \tau^{(m+n+2)} \\ H_m(e^{j\pi/4}\tau x) H_n(e^{j\pi/4}\tau y) \tag{8.20}$$

where $\tau = \sqrt{k/2(z + jc)}$ and H_n are the Hermite polynomials.

Figures 8.25 and 8.26 show the radiation pattern of the complex dipole p_{100} and quadrupole p_{110} for a $\vec{\beta}$ parallel to the principal axis X, Y, and Z. Figures 8.27 and 8.28 show a comparison of the radiation pattern between the exact expression (8.15) and the approximated expression (8.20).

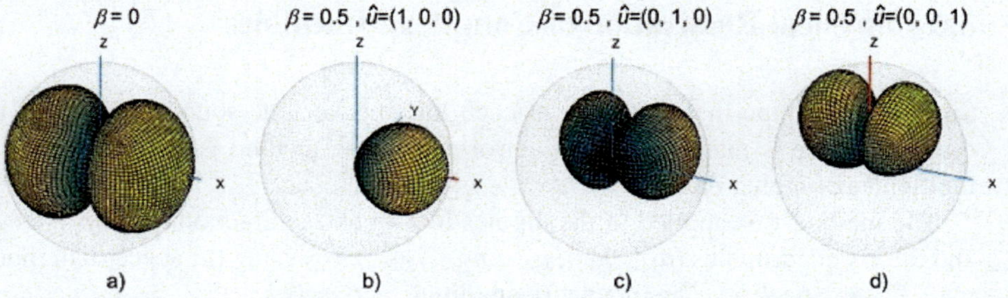

Fig. 8.25 Dipole p_{10} with $|\beta| = 0$ and $|\beta| = 0.5$

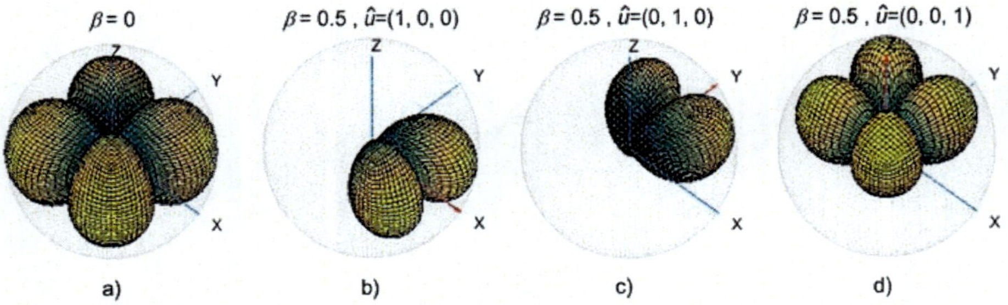

Fig. 8.26 Lateral quadrupole p_{11} with $|\beta| = 0$ and $|\beta| = 0.5$

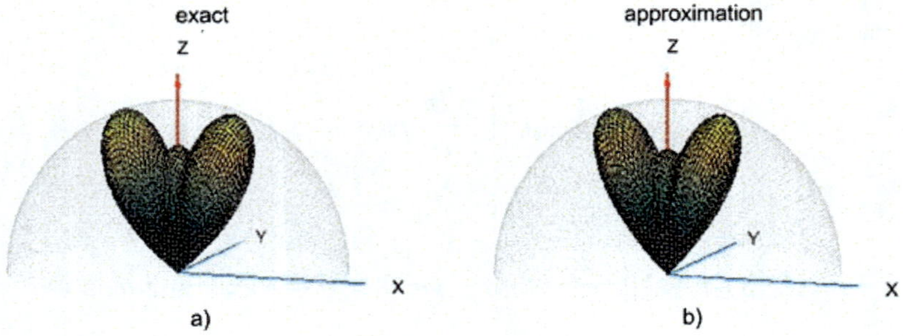

Fig. 8.27 Longitudinal quadrupole p_{20} with $\vec{\beta} = (0, 0, 3)$

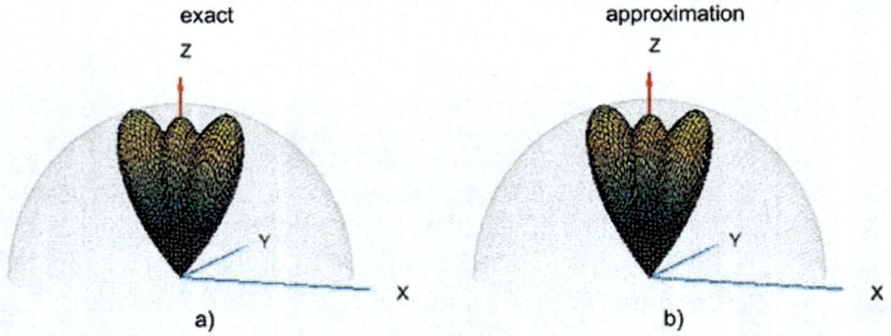

Fig. 8.28 Longitudinal quadrupole p_{20} with $\vec{\beta} = (0, 0, 6)$

8.6 Physical Realization of Complex Multipoles

Since to determine the amplitude of each loudspeaker, the sound pressure of the original source is matched at the control points, the method is also applicable to multipoles of higher orders.

The method was applied to the dipoles in *x*, *y*, and *z* direction ($p_{100}, p_{010}, p_{001}$) and the six quadrupoles ($p_{110}, p_{101}, p_{011}, p_{200}, p_{020}, p_{002}$) using the soccer ball model with 32 loudspeakers. The imaginary amplitude is $\beta = 0.2$.

Table 8.7 Relative error at the control points (ε_C) for multipoles of order 0, 1 and 2, $\beta = 1$ and $A_f = 1$ for the soccer ball model with 32 loudspeakers

	125 Hz	250 Hz	500 Hz	1000 Hz
p_{000}	**0.3**	**0.3**	**0.5**	**4.9**
p_{100}	**3.4**	**2.4**	**2.7**	*13.2*
p_{010}	**2.8**	**1.8**	**1.4**	**7.7**
p_{001}	**3.6**	**2.4**	**2.1**	**9.0**
p_{110}	*20.0*	*13.7*	**9.2**	*21.8*
p_{101}	*21.6*	*14.8*	**10.0**	*22.7*
p_{011}	*21.8*	*14.5*	**8.2**	*15.9*
p_{200}	*15.9*	**9.4**	**5.3**	*17.8*
p_{020}	*16.6*	**9.0**	**4.0**	*10.7*
p_{002}	*18.6*	*10.2*	**4.9**	*12.6*

Table 8.7 shows the relative error at the control points (ε_C). The results show that the sound radiation of the complex dipoles can be well reproduced with an error below 10% (except only p_{100} at 1000 Hz). The quadrupoles however have in the majority of the cases an error larger than 10% (but less than 25%). The error of the complex monopole rises with increasing frequency. On the contrary, the relative error of the higher order multipoles sinks with increasing frequency but for $f < 1000$ Hz. The reason for that behavior may lie in the size of the near field radiation, which is small for the complex monopole but increases with the order of the complex multipole. For the dipole, the factor responsible for the near field of the pressure is proportional to $1 + 1/jkr$ while for the quadrupole, it is proportional to $1 + 3/jkr + 3/(jkr)^2$. These factors are inversely proportional to the frequency, therefore, the size of the near field decreases with increasing frequency.

Figure 8.29 shows the variation of the near field factors around the soccer ball model for a dipole at 125 Hz and for a quadrupole at 125, 250, and 500 Hz. The reduction of the size of the near field is evident. One can observe that the control points are placed inside the near field at 125 and 250 Hz, but outside the near field at 500 Hz.

The relative error at points in the far field, ε_F, defined as in (8.14) was also computed. Table 8.8 shows the results of the simulation when the points were distributed on the surface of a sphere of radius $R = 10$ m. Interestingly, the error at the far field points is small even though the error at the control points is larger than 10%. This can be explained by the decay of the near field. Hence, the model can be used to reproduce the sound radiation of the complex multipoles in the far field.

To illustrate this effect, a spatially dependent local relative error $\varepsilon_p(\vec{r})$ was also defined:

$$\varepsilon_p(\vec{r}) = 100 \times \left| \frac{p^{(s)}(\vec{r}) - p^{(e)}(\vec{r})}{p^{(e)}(\vec{r})} \right| \tag{8.21}$$

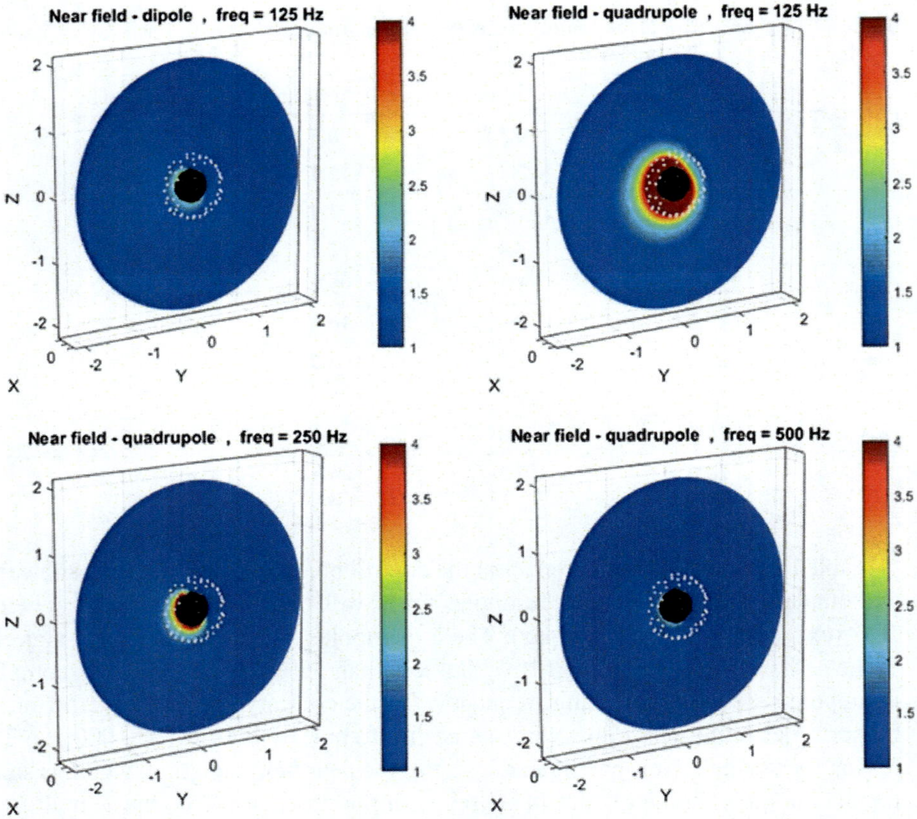

Fig. 8.29 Near field for a complex dipole and a complex quadrupole with $\beta = 0.2, A_f = 1$

Table 8.8 Relative error in the far field (ε_F) for multipoles of order 0, 1 and 2, $\beta = 1$ and $A_f = 1$ for the soccer ball model with 32 loudspeakers

	125 Hz	250 Hz	500 Hz	1000 Hz
p_{000}	**0.0**	**0.0**	**0.2**	**4.1**
p_{100}	**0.0**	**0.1**	**1.3**	*11.3*
p_{010}	**0.1**	**0.1**	**0.5**	**6.6**
p_{001}	**0.0**	**0.1**	**1.0**	**7.7**
p_{110}	**0.3**	**0.8**	**4.6**	*18.7*
p_{101}	**0.3**	**1.0**	**5.1**	*19.9*
p_{011}	**0.5**	**0.9**	**4.0**	*13.7*
p_{200}	**0.4**	**0.3**	**1.7**	*15.6*
p_{020}	**0.7**	**0.4**	**1.4**	**9.3**
p_{002}	**0.4**	**0.4**	**2.1**	*11.2*

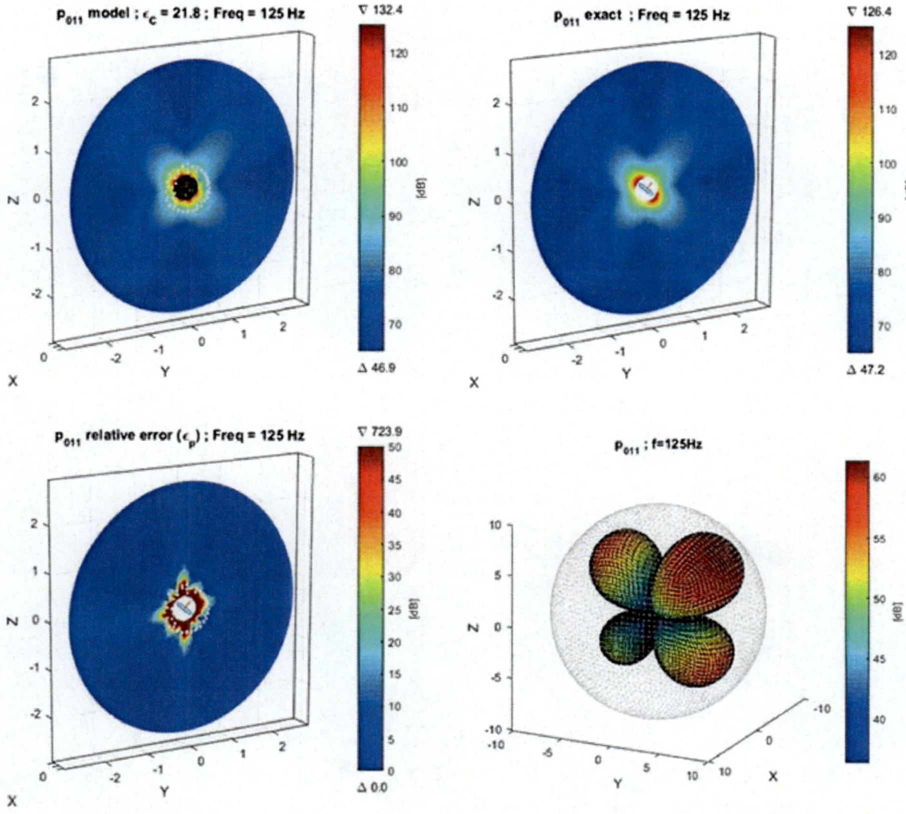

Fig. 8.30 Sound radiation from soccer ball model and analytical solution for the quadrupole p_{011} with $\beta = 0.2, A_f = 1$

Figure 8.30 shows simulated and exact sound radiation and the local relative error in the field points ε_p.

Near the model surface and around the control points ε_p can be very high, up to 700% but it decreases very fast in the space and becomes very small at some distance from the singularity.

Chapter 9
Simulation of Vibrating and Scattering Objects with ESM/CESM

9.1 Conventional Equivalent Source Method (ESM)

The equivalent source method (ESM) is a general tool for calculating the sound radiation or scattering from complex-shaped structures into the three-dimensional space. The basic idea of the method consists in replacing the structure by a system of acoustical sources placed in the interior of the structure. By definition, these source functions have to satisfy the Helmholtz equation and the radiation condition. For solving the radiation or scattering problem completely, the source system also has to fulfill the boundary conditions on the surface of the body or, equivalently, a certain boundary equation. The most general form of a local boundary condition is written as

$$R(p) = a\frac{\partial p}{\partial n} + bp = f, \tag{9.1}$$

where R is a linear operator, n is the normal vector of the body surface, and a, b, and f are known functions.

It is assumed that the pressure can be presented in the form

$$p(\vec{x}) = \sum_{i=1}^{N} c_i \psi_i(\vec{x}, \vec{y}), \tag{9.2}$$

where N is the number of sources, $\vec{x}$ is an arbitrary point in space, $\vec{y}$ is the position of the equivalent sources, ψ_i the source functions, and c_i the unknown coefficients.

For a given N, the better the system of sources satisfies the boundary condition on the surface of the structure, the closer is the agreement between the original and the simulated sound field. The coefficients are obtained by minimizing the boundary error ε defined as

$$\varepsilon(\vec{x}) = R(p) - f(\vec{x}) \tag{9.3}$$

M. Ochmann and R. Piscoya, *Theory and Application of Acoustic Sources Using Complex Analysis*, https://doi.org/10.1007/978-981-33-6040-2_9

The minimization process can be performed by choosing a complete family of weighting functions w_m; $m = 1, 2, 3, \ldots$ and demanding that

$$\int_S \varepsilon(\vec{x}) w_m(\vec{x}) \mathrm{d}^2 x = 0 \tag{9.4}$$

There are many options for w_m but frequently the least square minimization technique is used which requires $w_m = [R(\psi_m)]^*$. This choice implies that (9.4) can be written as

$$\frac{\partial}{\partial c_m} \int_S \left| \sum_{i=1}^{N} c_i R(\psi_i(\vec{x})) - f(\vec{x}) \right|^2 \mathrm{d}^2 x = 0 \quad m = 1, \ldots, N \tag{9.5}$$

By performing (9.5), a system of N equations with N unknowns is obtained, which in matrix form can be written as

$$Ac = d \tag{9.6}$$

where A is a square Hermitian matrix, c is the vector of unknowns which determine the source strengths, and d is the right side vector

$$A_{mn} = A_{nm}^* = \int_S [R(\psi_m)]^* R(\psi_n) \mathrm{d}^2 x \text{ and } d_m = \int_S f[R(\psi_m)]^* \mathrm{d}^2 x \tag{9.7}$$

The spherical wave functions present the simplest form of radiating functions. Hence, they are the type of sources most often used, and the pressure (9.2) can be written as

$$\sum_{i=1}^{N} \sum_{n=0}^{Q} h_n^{(2)}(k|\vec{r} - \vec{r}_i|) \sum_{m=0}^{n} P_n^m(\cos\gamma_i)(A_{nm}\cos(m\delta_i) + B_{nm}\sin(m\delta_i)) \tag{9.8}$$

where Q is the order of the multipole, $\cos(\gamma_i) = \frac{z - z_i}{|\vec{r} - \vec{r}_i|}$ and $\tan(\delta_i) = \frac{y - y_i}{x - x_i}$.

In practical applications, the important question arises of how to find optimal number, type, and locations of the sources in the interior of the structure. Clearly, the main requirement is that these parameters have to be chosen such that the boundary error becomes minimal but there is no general criterion to choose them. It depends on the type of problem being investigated. In general, the position of the equivalent sources is important when monopoles (zero order) are used, but when sources of higher orders like dipoles or quadrupoles are considered, a uniform distribution of the sources inside the control surface is sufficient to provide reasonable results.

The accuracy of the results increases with the number of equivalent sources, but this number cannot be indefinitely raised because the system becomes unstable. The stability of a system of equations like (9.6) can be investigated by considering the condition number

$$\kappa = \mathrm{cond}(A) = \|A\| \left\|A^{-1}\right\| \tag{9.9}$$

where $\|\ \|$ is a matrix norm and κ is a real number greater than or equal to one. The condition number measures the sensitivity of the solution c with respect to perturbations in A or in d. An unstable or badly conditioned system has a large condition number. In this case, a regularization method has to be employed to obtain an approximate solution. Two methods that are frequently used are the truncated singular value decomposition (TSVD) and the Tikhonov regularization, which define the solution vector in terms of the eigenvalues of the matrix.

The application of the ESM will be shown in the following numerical examples.

9.1.1 Sound Radiation of a Rectangular Radiator

A rectangular parallelepiped with dimensions 0.25 m × 0.275 m × 0.3 m vibrates with the normal velocity equal to the normal velocity produced by an oscillating sphere with radius $R_s = 0.03$ m and center at $r_q = (0.15\text{ m}, 0.165\text{ m}, 0.18\text{ m})$. A sphere oscillating with velocity v_0 parallel to the z_s-axis has a normal velocity $v_0\cos\theta$, where θ is the angle with the z-axis on a coordinate system with origin at the center of the sphere. The sphere produces a sound pressure and particle velocity

$$p = \rho c v_0 A\left(\frac{R_s}{r}\right)\cos\theta\left(1 + \frac{1}{jkr}\right)\mathrm{e}^{-jk(r-R_s)} \tag{9.10}$$

$$\begin{aligned} v_r &= v_0 A\left(\frac{R_s}{r}\right)\cos\theta\left(1 + \frac{2}{jkr} + \frac{2}{(jkr)^2}\right)\mathrm{e}^{-jk(r-R_s)} \\ v_\theta &= v_0 A\left(\frac{R_s}{r}\right)\sin\theta\left(\frac{1}{jkr} + \frac{1}{(jkr)^2}\right)\mathrm{e}^{-jk(r-R_s)} \end{aligned} \tag{9.11}$$

with $A = \left(1 + \frac{2}{jkR_s} + \frac{2}{(jkR_s)^2}\right)^{-1}$.

The sound power of the oscillating sphere is given by

$$W = \frac{2}{3}\rho c\pi R_s^2 v_0^2 \frac{(kR_s)^4}{4 + (kR_s)^4} \tag{9.12}$$

To determine the radiated sound of the box using the ESM, six source positions are considered as shown in Fig. 9.1. At each position, three set of multipoles will be considered: (a) 1 monopole, (b) 1 monopole + 3 dipoles and (c) 1 monopole + 3 dipoles + 5 quadrupoles. To evaluate the accuracy of the simulated results, the relative surface error for the different multipoles sets is compared. Additionally, two

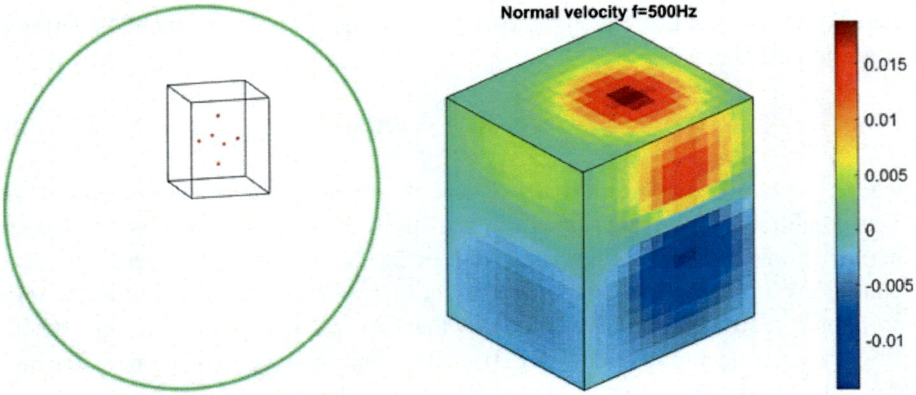

Fig. 9.1 *Left*: Position of equivalent sources (red points) and field points (green points); *right*: distribution of normal velocity on the surface

other errors are considered, error in the sound power W_s and error in the pressure at a set of field points p_F:

$$\varepsilon_W = \frac{|W_s - W|}{W} \tag{9.13}$$

$$\varepsilon_{fp} = \sqrt{\frac{\sum_{i=1}^{M} \left|p_i - p_{F,i}\right|^2}{\sum_{i=1}^{M} |p_i|^2}} \tag{9.14}$$

where M is the number of field points. The stability of the solutions are evaluated with the condition number.

The surface error decreases by increasing the order of the multipoles, it varies between 30 and 50% using only monopoles, between 10 and 20% using monopoles and dipoles and below 10% using monopoles, dipoles, and quadrupoles. Inversely, the condition number grows by increasing the order of the multipoles (see Fig. 9.2).

The sound power has a small relative error, below 10% for the three configurations, while the sound pressure at the field points remains below 10% for configuration (b) and (c) (see Fig. 9.3). The error in the pressure at the field points is smaller than the error at the surface, since at the surface the near field of the equivalent sources is still important.

9.1.2 Sound Scattering of a Non-convex Cat's Eye

The model of the cat's eye consists of 7911 elements. The radius of the corresponding sphere is 0.4 m so that a minimum of six elements per wavelength is achieved for frequencies up to 3000 Hz.

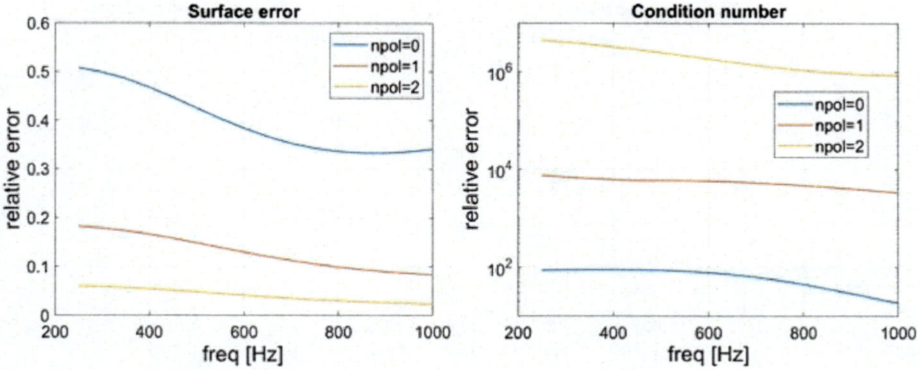

Fig. 9.2 Surface error and condition number

The incident wave is a plane wave with amplitude p_0 and an incident direction $(\theta_0, \phi_0) = (3\pi/4, 3\pi/2)$. The expressions for the sound pressure and particle velocity are:

$$p_{\text{inc}} = p_0 \mathrm{e}^{-j\vec{k}_{\text{inc}} \cdot \vec{r}}, \vec{k}_{\text{inc}} = k_0 \hat{n}_{\text{inc}}$$

$$\vec{v} = \hat{n}_{\text{inc}} \frac{p_{\text{inc}}}{\rho c}$$

where $\hat{n}_{\text{inc}} = (\sin\theta_0 \cos\phi_0, \sin\theta_0 \sin\phi_0, \cos\theta_0)$ and k_0 is the wave number.

The scattering problem is treated as a radiation problem with the normal velocity of the structure equal to $-v_{\text{ns}}$.

In a first calculation, multipoles of order 9 were placed in the middle of each of the seven octants. A total of 700 equivalent sources are used to reproduce the velocity distribution at the surface.

With this equivalent source distribution, a relative error smaller than 20% is obtained for frequencies below 3000 Hz. For frequencies below 1000 Hz, the condition number is bigger than 10^{16} and a regularization method has to be used to solve the matrices (see Fig. 9.4).

In Fig. 9.5, the radiation patterns for two different frequencies are shown. It can be seen that the forward scattering and the backscattering are good modeled.

A second equivalent source distribution with more source positions and lower order of the multipoles was tested. 42 points inside the cat's eye over a sphere of radius 0.24 m with angles $\theta = 15, 45, 75, 105, 135$ and $\phi = 22.5, 67.5, 112.5, 157.5, 202.5, 247.5, 292.5, 337.5$ were selected. Multipoles of second order were used making a total of 378 sources. The results with the second distribution are shown in Figs. 9.6 and 9.7. The surface error is very similar to the first case in the low frequencies. Beyond 2500 Hz, the surface error increases faster since the number of total sources is smaller, almost a half of the sources in the first case. The condition number is of the same order. Since the surface errors are very similar in both cases, similar

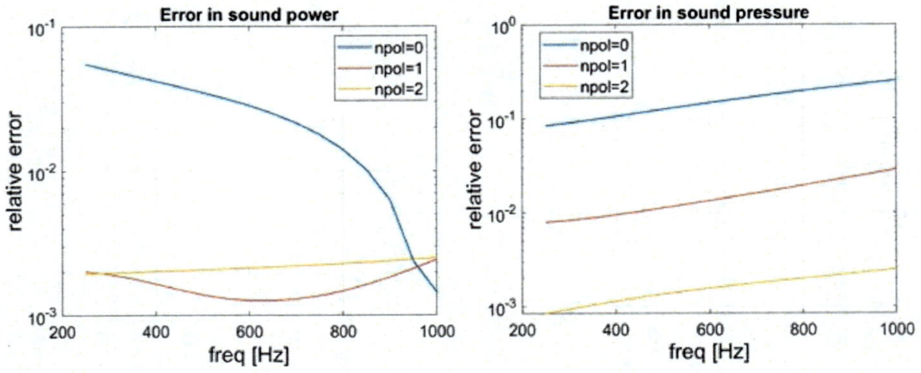

Fig. 9.3 Relative errors in sound power and sound pressure

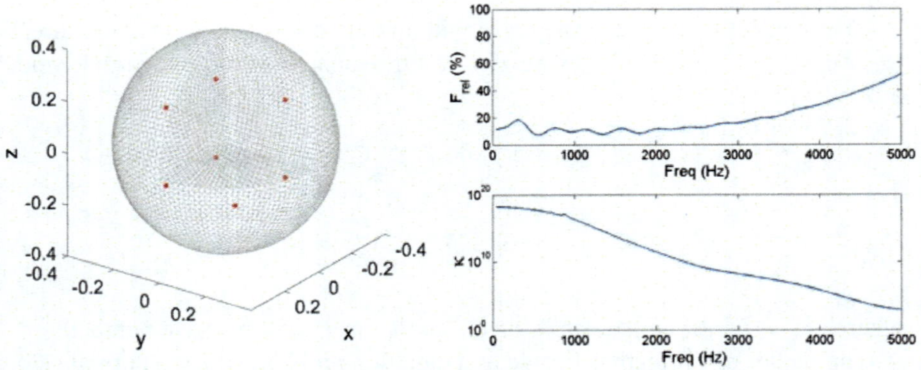

Fig. 9.4 Position of the equivalent sources (left) and the variation of the surface error F_{rel} and condition number κ with the frequency (right)

radiation patterns are expected. A comparison of the radiations patterns of Figs. 9.5 and 9.7 confirms that. The radiation pattern at 2840 Hz is symmetric with respect to the axis of the incident wave with the first source distribution. This symmetry is less pronounced with the second source distribution.

9.2 Complex Equivalent Source Method (CESM)

In its traditional form, the ESM only uses sources which have real source positions ("real sources"). Now, the ESM is extended by adding sources with complex source positions ("complex sources"). We denote such an enhanced method as complex equivalent source method and use the abbreviation CESM. Thus, the CESM replaces the vibrating structure by a system of real and complex sources in such a way that

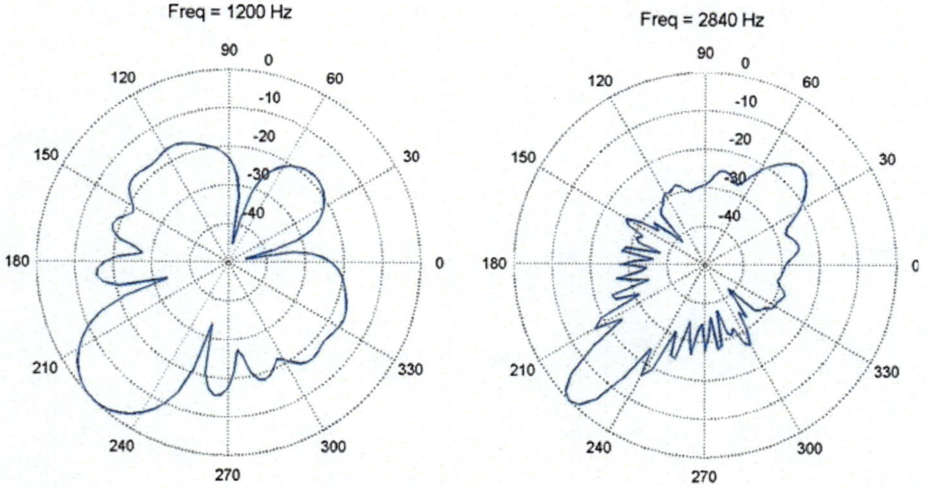

Fig. 9.5 Radiation patterns of the scattered sound for two frequencies

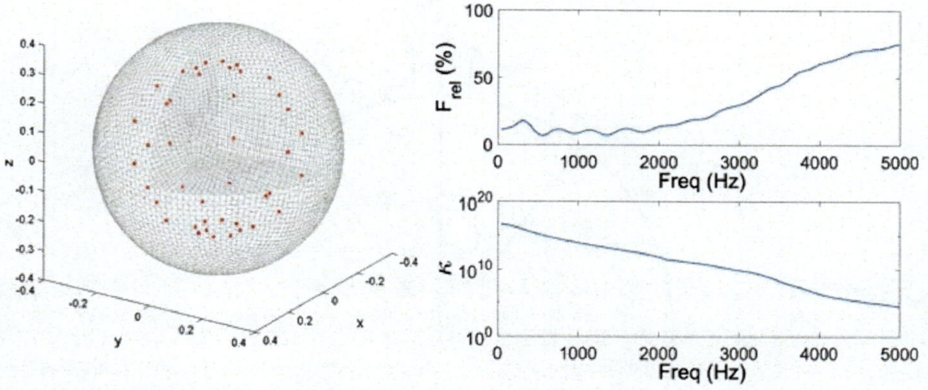

Fig. 9.6 Position of the equivalent sources (left) and the variation of the surface error F_{rel} and condition number κ with the frequency (right)

the sound field radiated or scattered by the structure is approximated by the source system.

Since the extended source system fulfills the Helmholtz equation and the radiation condition automatically, it is only necessary that the source system satisfies the boundary condition at the structural surface S, in order to obtain an accurate solution of the problem. As in the ESM, the amplitude of the equivalent complex monopoles is determined by minimizing the error on the boundary.

Like in the ESM, the positioning of the equivalent sources is also an important issue for the CESM. For the usual ESM, equivalent sources should be placed at a certain distance from the surface of the structure, but always inside it. The fact that the singularity of the complex monopole is not only a point but a disc of radius β, introduces a limit to the values of the complex part of the source position that can be

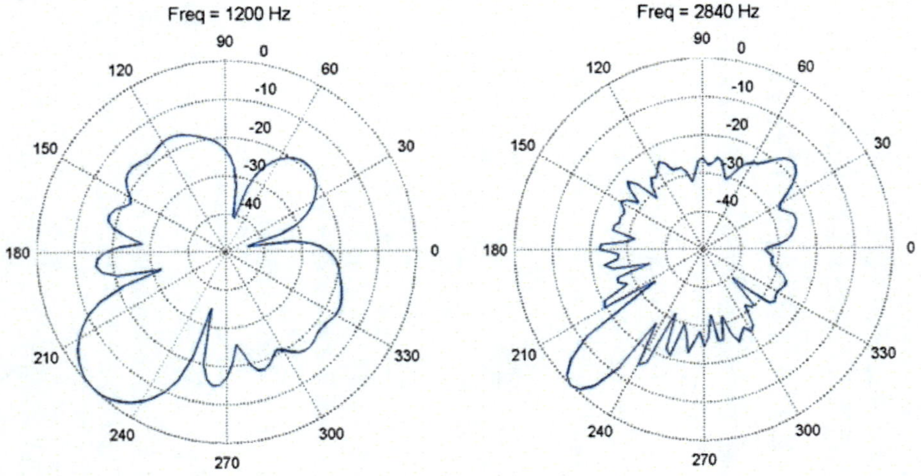

Fig. 9.7 Radiation patterns of the scattered sound for two frequencies

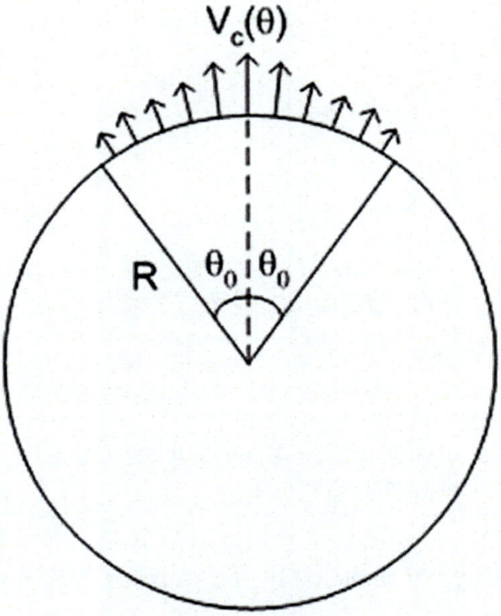

Fig. 9.8 Oscillating spherical cap

chosen. In this sense, the positioning of a complex monopole requires more attention than in the case of real multipoles.

The application of the CESM will be shown in the following numerical examples.

9.2.1 Sound Radiation of a Spherical Cap

A simple radiator is represented by a curved piston embedded in a hard sphere. Using spherical coordinates, the normal velocity at the surface can be written as (Fig. 9.8):

$$V_n = \begin{cases} V_c(\theta) & 0 < \theta < \theta_0 \\ 0 & \theta_0 < \theta < \pi \end{cases} \tag{9.15}$$

If we consider the following solution for the sound pressure:

$$p(r, \theta) = \sum_n A_n h_n^{(2)}(kr) P_n(\cos\theta) \tag{9.16}$$

impose the boundary condition (9.15) and use the orthogonality relations of the Legendre Polynomials, the coefficients A_n are found to be:

$$A_n = \frac{\rho c}{j N_n h_n^{'(2)}(kR)} \int_0^{\theta_0} V_c(\theta) P_n(\cos\theta) \sin\theta \mathrm{d}\theta, \; N_n = \frac{2}{2n+1} \tag{9.17}$$

When V_c is constant $V_c(\theta) = V_0$, the expression for the coefficients is given by

$$A_n = \frac{-j\rho c}{N_n h_n^{'(2)}(kR)} T_n, \; T_n = \frac{V_0}{2} N_n (P_{n-1}(\cos\theta) - P_{n+1}(\cos\theta)) \tag{9.18}$$

and the sound power of the oscillating cap has a simple expression:

$$P = \frac{\pi}{\rho c k^2} \sum_n |A_n|^2 N_n \tag{9.19}$$

We study a cap of $\theta_0 = 30°$ in a sphere of radius $R = 0.25$ m for a frequency range that goes from 100 to 5000 Hz. For these parameters, the total number of terms used for the analytical solution is 40, since at 5000 Hz the amplitudes of the terms for $n >$ 30 are very small. For lower frequencies, the terms could be cut at a lower number.

For the numeric determination of the sound field, a model of a sphere with 5400 elements was generated. With this model, the condition of six elements pro wavelength is assured for frequencies under 5000 Hz. All elements whose centers lie within an angle range below 30° oscillate with a normal velocity of 1 m/s. The other elements are at rest. For the evaluation of the accuracy of the numeric simulations, the sound power and the sound pressure at points over the radiation axis and at points over a sphere of radius 8 m (32R) are compared.

As shown in Fig. 9.9 (picture on the top right), the analytical result (9.19) for the sound power level agrees well with the results obtained from the CESM and ESM. The surface error obtained with the complex sources is bigger than the error with the

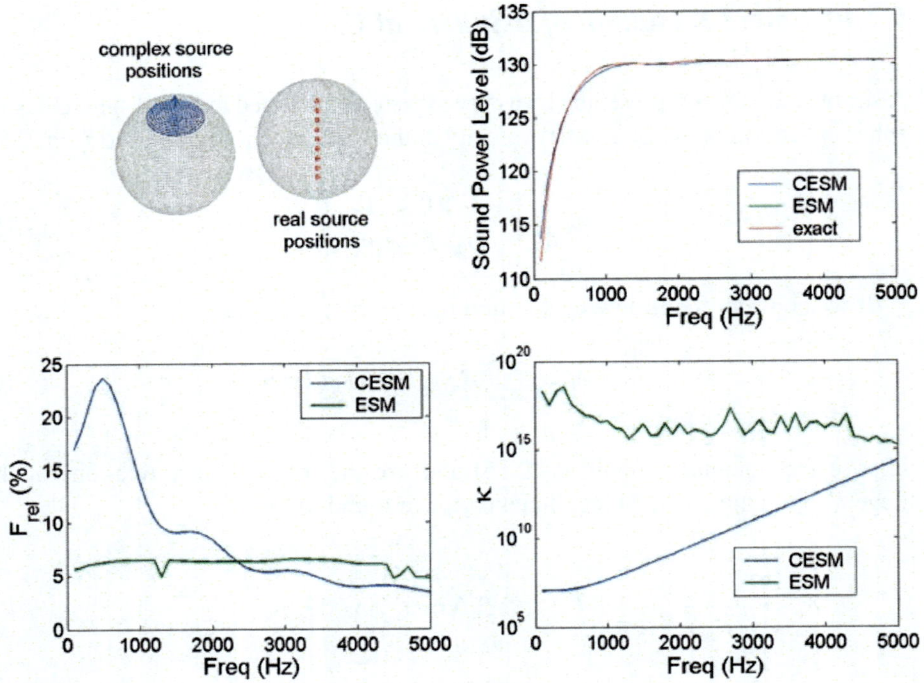

Fig. 9.9 Source distribution and results of the CESM and ESM

real sources at frequencies below 2500 Hz and slightly smaller for higher frequencies (see pictures at the bottom of Fig. 9.9). The real sources provide an almost constant surface error over the whole frequency range. With both type of sources, the error remains within an acceptable range. The condition numbers coming from the real sources are several orders of magnitude bigger than those of the complex sources. For these high values, a regularization method was required.

In Fig. 9.10, a comparison of the simulated and exact sound fields is made at points at the axis of the sphere (top) and in the far field (bottom). The agreement between simulated and analytical values is very good for both methods. At the axis, the CESM slightly underestimates the maximum at 2 kHz, while the ESM shows a bigger error at 4 kHz.

In the far field, the ESM reproduces the sound field almost perfectly while the CESM underestimates the radiation backward because the complex monopoles have a marked directivity in one half-space. The difference in computation time is remarkable, 14 s for the complex sources, 276 s for the real sources.

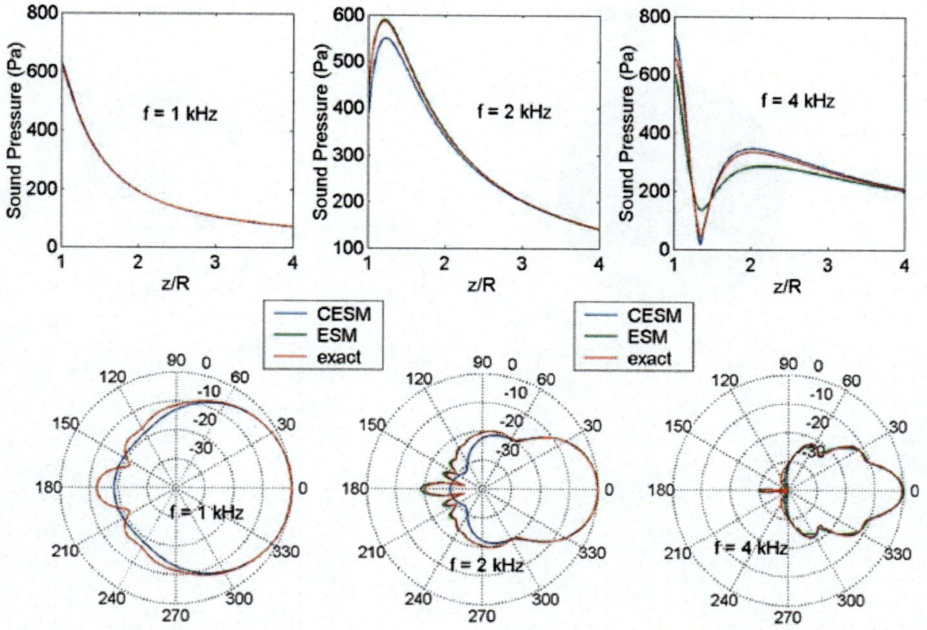

Fig. 9.10 Comparison of the sound pressure at the axis (top) and in the far field (bottom) obtained by the CESM and ESM at three different frequencies

9.2.2 Sound Scattering of a Cat's Eye

Three different source point distributions (CSD1, CSD2, and CSD3) were taken to make the calculations (Fig. 9.11). All three configurations have a group of five complex monopoles pointing in the direction of the incident wave which should model the forward scattering. CSD1 and CSD2 have complex sources behind the perpendicular planes that define the structure. CSD3 has a group of five complex sources pointing at the opposite direction of the plane wave.

The surface error obtained with these three configurations can be seen in Fig. 9.12. The errors are similar and remain over 40% for the all the frequencies. The condition numbers increase with the frequency as expected but do not surpasses 10^{13}. The radiation patterns at 1200 Hz and 2840 Hz are compared in Fig. 9.13. The main lobe of the scattered wave is the same with the three configurations but rather wider than the one obtained with the real sources. Differences appear for the other angles. The patterns are symmetric with respect to the incident wave axis. The radiation of the CSD3 shows a more pronounced backscattering, especially at 2840 Hz, which is expected since the equivalent sources have been positioned pointing at that direction.

In order to reduce the surface error, real monopoles were added to the complex sources distributions. The same 42 positions used with the ESM (Sect. 9.1.2) were filled with monopoles. The surface error was reduced to 20% in the frequency range of 0–1500 Hz but for higher frequencies the real monopoles do not bring any improvement (Fig. 9.14) low small differences between 5° and 90° and are not anymore symmetric.

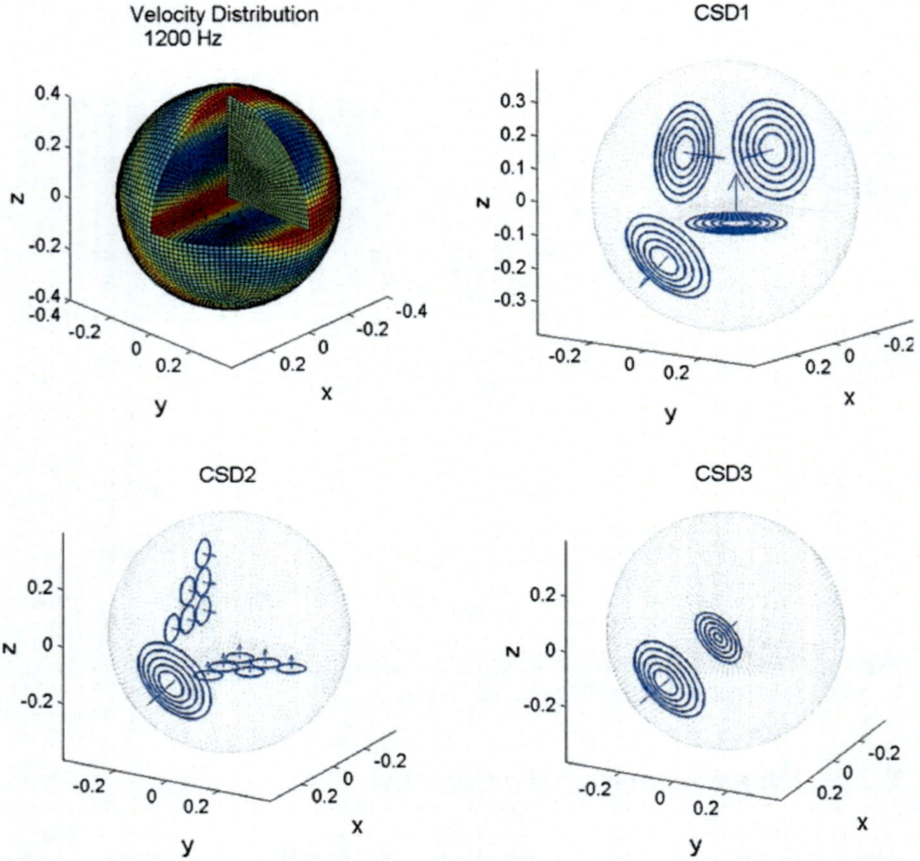

Fig. 9.11 Different distributions of complex source positions

9.2.3 *Sound Radiation of a Non-convex Cat's Eye*

For the modeling of sound radiation, it is assumed that the interior of each of the three perpendicular plane surfaces of the structure vibrates with a constant velocity of amplitude 1 m/s (Fig. 9.15). The sound field is again simulated with real multipoles and complex monopoles.

The calculations with only real multipoles are made using the same two distributions of sound positions of Sect. 9.1.2, 7 points and 42 points. The curves of relative error and condition number for these distributions are shown in Fig. 9.16. In the legend, the first number indicates the number of source positions, the second the order of the multipole. The errors are big in the low frequencies up to 500 Hz and then decrease to 20%. At high frequencies, the errors increase with the frequency. Increasing the order of the multipoles elevates the ill-conditioning of the matrices. The dotted line in the plot of the condition numbers marks the limit 10^{16}. Above this value, regularization methods are needed. Reasonable results can be expected

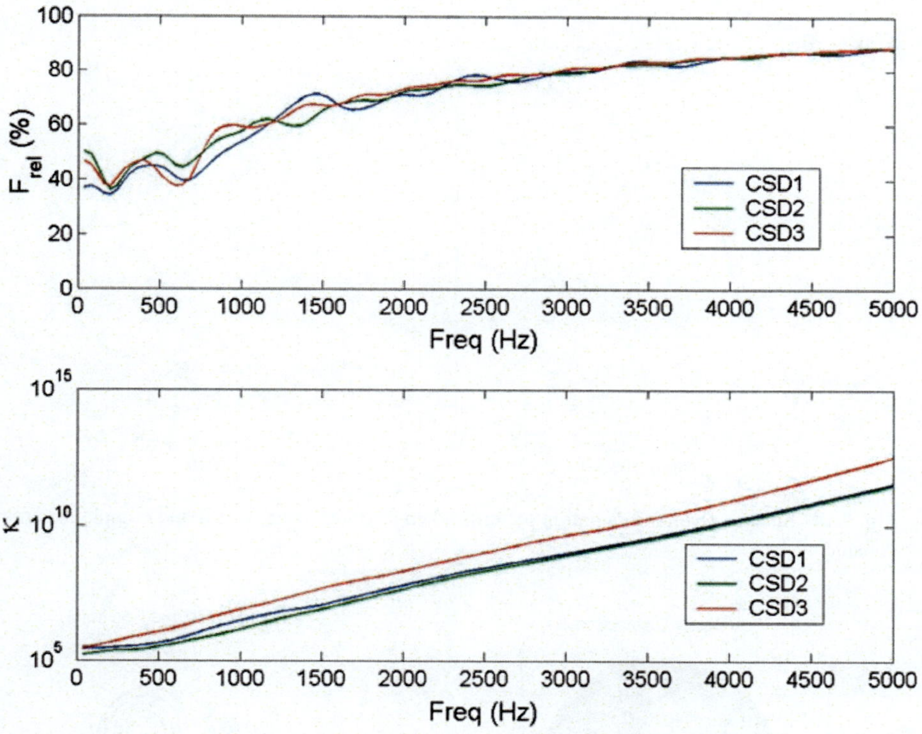

Fig. 9.12 Comparison of the F_{rel} and K between CSD1, CSD2, and CSD3

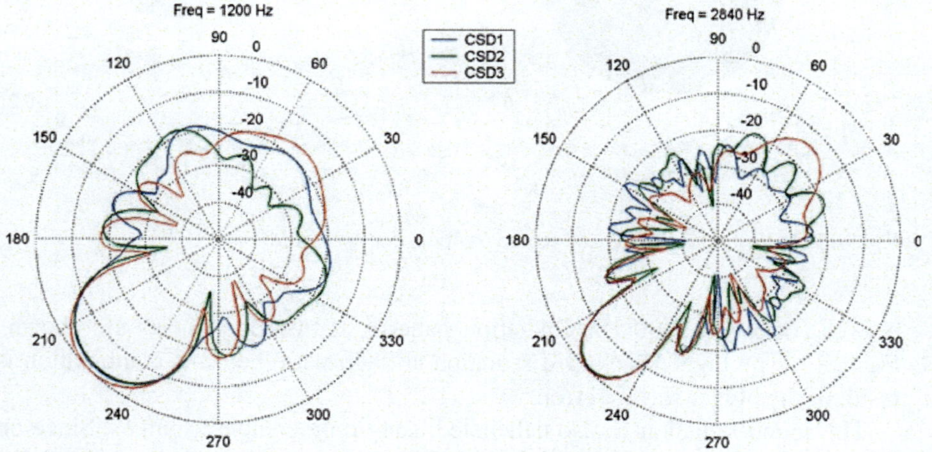

Fig. 9.13 Comparison of the radiation patterns of CSD1, CSD2, and CSD3

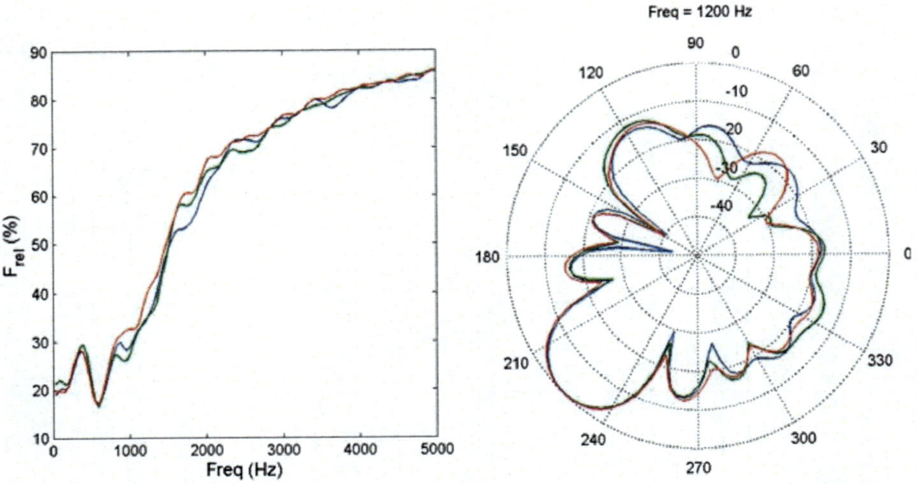

Fig. 9.14 Surface error and radiation patterns when a combination of real and complex sources is considered

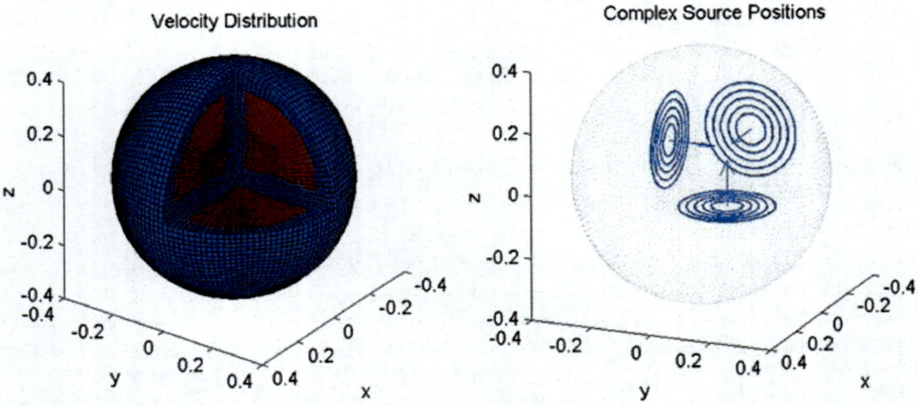

Fig. 9.15 Velocity distribution and source positions for the CESM

between 500 and 3000 Hz. Radiation patterns at two frequencies are shown in Fig. 9.17. The bigger backward radiation at 2840 with the 42–2 combination is a result of the bigger surface error.

The sound radiation is also calculated using only complex sources. Since only the plane surfaces vibrate, it is reasonable to put the complex sources behind them, as is shown in Fig. 9.15. A second distribution of complex sources is conceived adding five complex monopoles near the middle of the structure pointing at the (1, 1, 1) direction. These two distributions are used for the calculation of the sound field and also combined with the group of 42 real monopoles as in the scattering case. The surface errors and condition numbers are shown in Fig. 9.18. In the legend, the first number corresponds to the number of complex monopoles and the second to the number of real monopoles. It can be seen that the complex sources provide

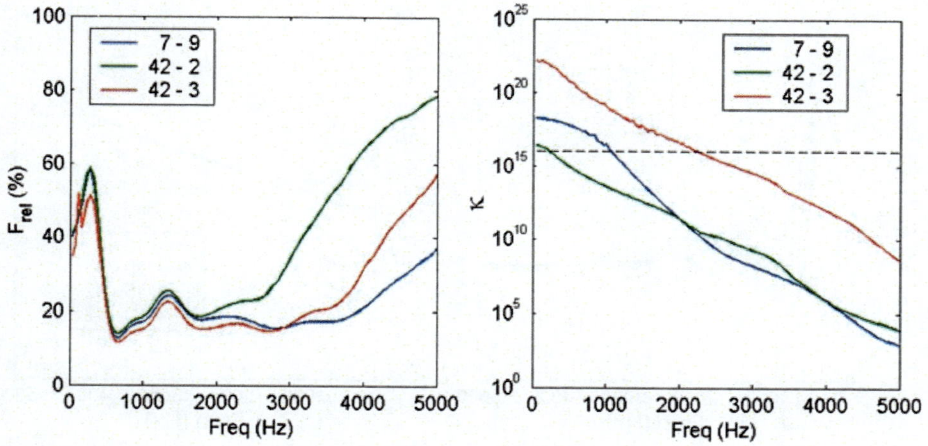

Fig. 9.16 Relative error and condition number using real multipoles

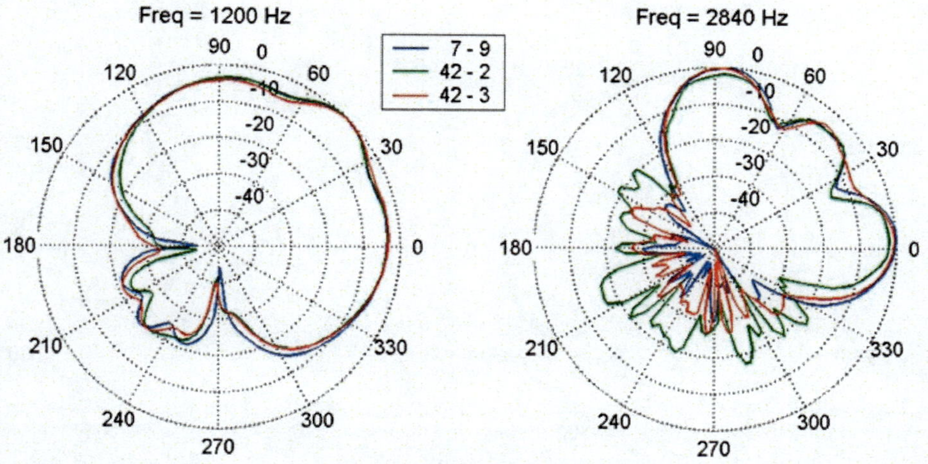

Fig. 9.17 Radiation patterns of the radiating cat's eye

bigger errors than the multipoles at low frequencies but they do not increase with the frequency, like those of the multipoles. Again, the addition of real sources reduces the surface error especially at low frequencies and alters the symmetry of the radiation patterns. At 2840 Hz, the complex sources produce two main lobes at 0° and 90° and a small lobe at 45° (Fig. 9.19). The multipoles give instead a lobe at 45° that is only slightly smaller than the other two (Fig. 9.17).

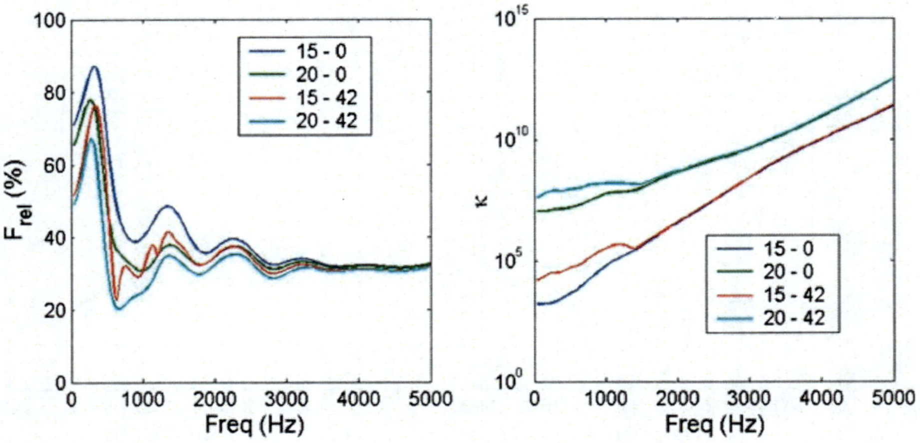

Fig. 9.18 Surface error and condition number of the complex source distributions

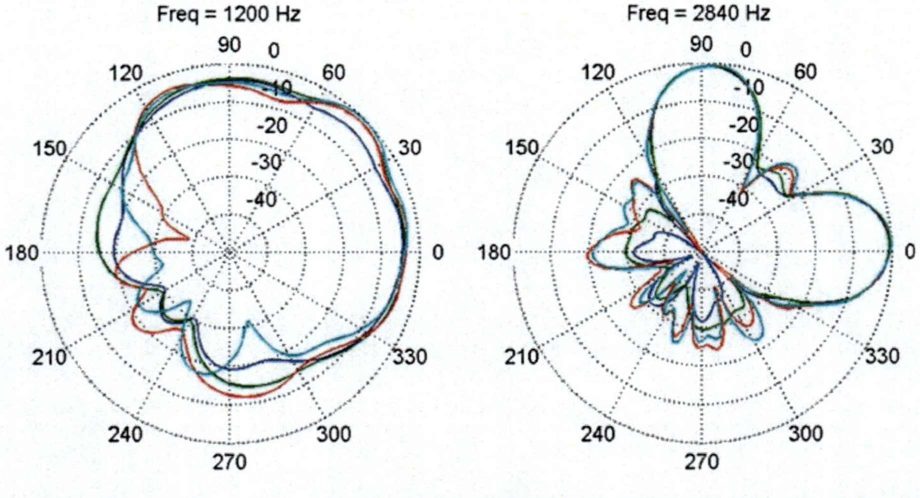

Fig. 9.19 Comparison of the radiation patterns obtained with the different complex source distributions

9.2.4 *Sound Radiation of a Vibrating Cone Embedded in a Baffle*

This case considers a vibrating cone embedded in an infinite rigid plane (baffle). If the vibrating surface were flat and placed on the infinite plane, the radiated sound could be obtained easily with the Rayleigh integral. Since the vibrating surface is not flat and some volume below the plane is connected to the upper half-space, the problem becomes more complicated. The cone is small with a bigger radius of 60 mm and a smaller radius of 20 mm. The sound pressure will be calculated in the frequency range of [3000, 6000] Hz. Since the wavelength of the sound in the range

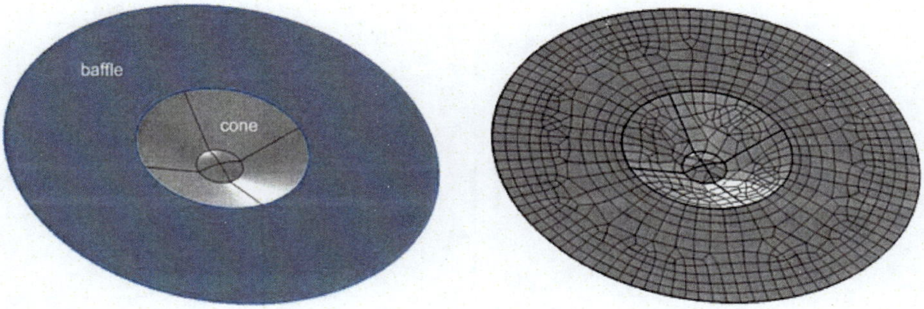

Fig. 9.20 Model of the cone embedded in a rigid baffle

of frequencies under study is smaller than the cone, certain focusing of the radiated pressure is expected. Due to this focusing, the simulation of the sound radiation applying the CESM is expected to provide accurate results. For the calculation, a discretized model is needed. The model should include the cone but also part of the rigid baffle. Figure 9.20 shows the geometry of the model and the mesh.

One calculation is performed using the conventional ESM. The sound radiation is obtained for 7 frequencies between 3000 and 6000 Hz. Figure 9.21 shows the positions of equivalent sources . The 37 positions are located below the cone and the baffle and distributed more or less uniformly within the model surface. First, only monopoles are considered, then monopoles and dipoles and finally monopoles, dipoles, and quadrupoles. The total numbers of sources are 37, 148, and 333, respectively.

From the results shown in Fig. 9.22, one can observe that an error below 10% cannot be achieved with monopoles alone. The error stays between 28 and 55%. With monopoles and dipoles, the error decreases and can be considered acceptable. With monopoles, dipoles, and quadrupoles, the error is further reduced. The condition number is large but remains in all cases in manageable values.

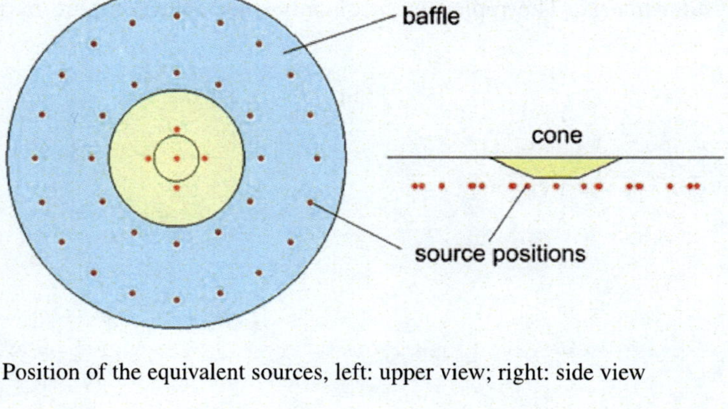

Fig. 9.21 Position of the equivalent sources, left: upper view; right: side view

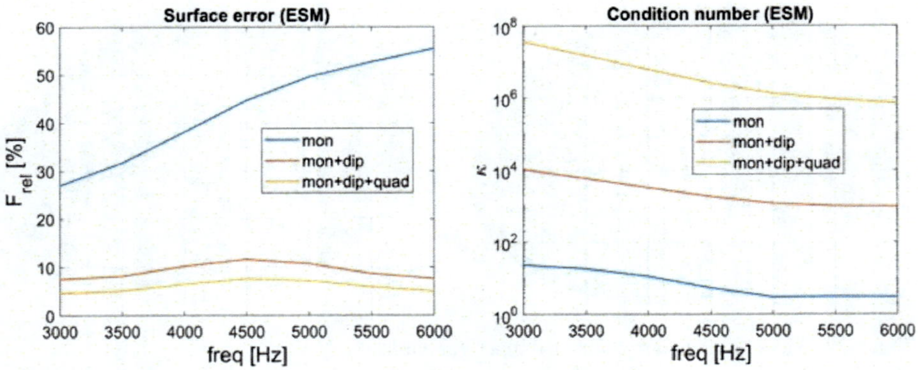

Fig. 9.22 Surface error and condition number for different source configurations (ESM)

A second calculation is performed applying the CESM. We consider the same source distribution of the first calculation. In the first test (a), the real monopole placed at the axis below the cone is replaced by a complex monopole with imaginary part $\vec{\beta} = (0, 0, 0.1)$. In a second test (b), five real monopoles are replaced by five complex monopoles with imaginary part $\vec{\beta} = (0, 0, 0.1)$. In the last test (c), five real monopoles are replaced by five complex monopoles, all located at the axis, with imaginary part given by $\vec{\beta} = (0, 0, b)$, with $b = 0.05, 0.0625, 0.075, 0.0875, 0.1$, respectively. The three different configurations are shown in Fig. 9.23. The real monopoles are illustrated with red circles and the complex monopoles with green circles.

Figure 9.24 shows the surface error and condition number of the system matrix for the considered real and complex sources configurations.

We observe that the use of complex monopoles can reduce the error in the surface without increasing the number of sources. The accuracy of the calculation with 37 monopoles, 5 from them complex and the rest real, is almost the same as the accuracy from 333 real sources, monopoles, dipoles, and quadrupoles. The drawback of the insertion of complex monopoles is the significantly increase in the condition number of the system matrix. The replacement of only one real monopole by a complex

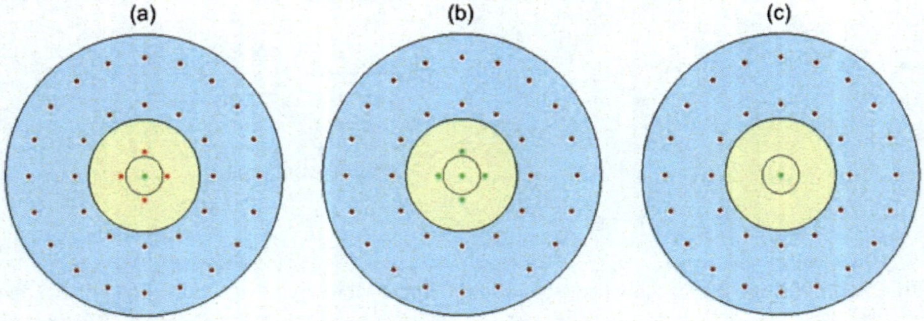

Fig. 9.23 Different source configurations of real and complex sources (CESM)

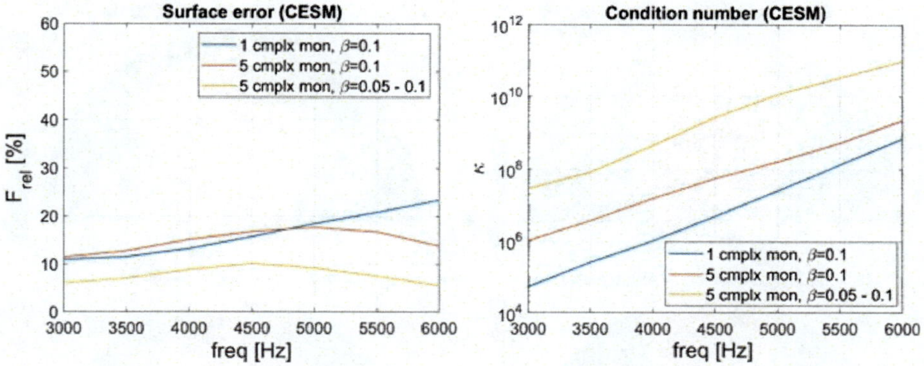

Fig. 9.24 Surface error and condition number for different source configurations (ESM)

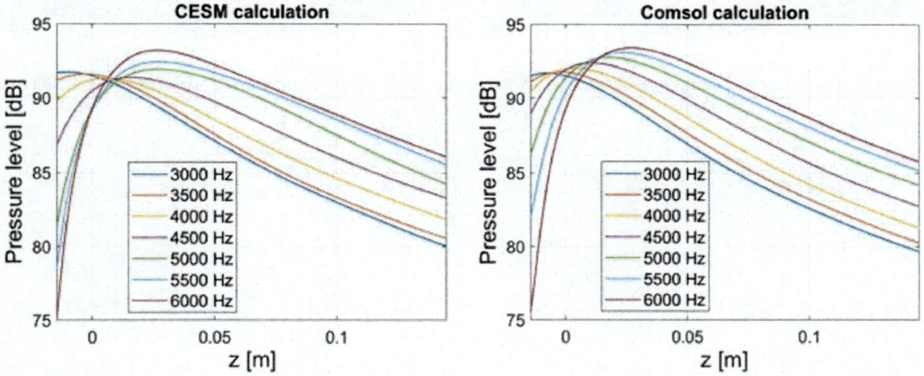

Fig. 9.25 Pressure level along the axis of the embedded cone obtained with the CESM and Comsol

one [case (a)] increases the condition number from less than 100 to about 15,000 at 3000 Hz and from 50 to about 10^9 at 6000 Hz. In all cases computed by the CESM, the values of condition number, although very large, were several orders below the maximum value at which a regularization method is needed (a reasonable upper limit is the inverse of the double-precision floating point relative accuracy, *eps* in Matlab).

Figure 9.25 shows a comparison of the results of the sound pressure level at several points on the axis of the cone. The agreement is very good. The differences are bigger near the cone surface and reduce with increasing distance. Figure 9.26 shows the pressure level distribution in a plane in front of the embedded cone at the frequencies 3000 Hz and 6000 Hz obtained with the CESM and with the reference with Comsol. With the much faster CESM is possible to obtain similar accuracy as with the FEM.

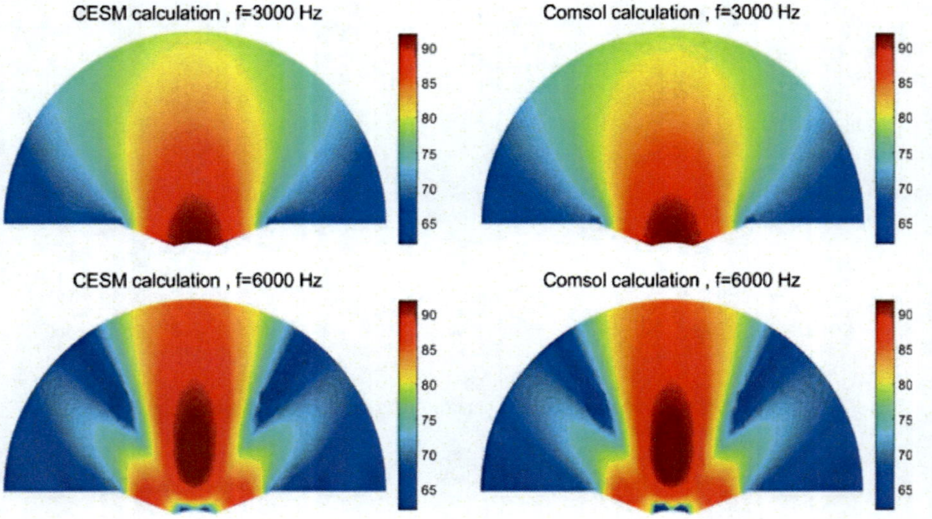

Fig. 9.26 Pressure level on a plane normal to the embedded cone obtained with the CESM and Comsol

Chapter 10
Green's Function Above Homogeneous Ground

10.1 Two-Dimensional Problem

Two semi-infinite lossless homogeneous media with parameters ρ_1, c_1 and ρ_2, c_2, respectively, with interface at $z = 0$ are considered. A line source is placed at the position (x_q, h) and a field point is located at position (x, z) (Fig. 10.1).

10.1.1 *Impulsive Time Dependence*

The differential equations for the velocity potential ϕ are given by

$$\frac{1}{c_1^2}\frac{\partial^2 \phi_I}{\partial t^2} - \Delta\phi_I = \delta(x - x_q)\delta(z - h)\delta(t - t_q), \quad z > 0, \tag{10.1}$$

$$\frac{1}{c_2^2}\frac{\partial^2 \phi_{II}}{\partial t^2} - \Delta\phi_{II} = 0, \quad z < 0, \tag{10.2}$$

with continuity of sound pressure and normal velocity at the interface as boundary conditions

$$-\rho_1 \frac{\partial \phi_I}{\partial t}\bigg|_{z=0} = -\rho_2 \frac{\partial \phi_{II}}{\partial t}\bigg|_{z=0}, \tag{10.3}$$

$$\frac{\partial \phi_I}{\partial z}\bigg|_{z=0} = \frac{\partial \phi_{II}}{\partial z}\bigg|_{z=0}. \tag{10.4}$$

The potential ϕ should satisfy the initial conditions

$$\phi_I(t \le t_q) = \phi_{II}(t \le t_q) = \frac{\partial \phi_I}{\partial t}(t \le t_q) = \frac{\partial \phi_{II}}{\partial t}(t \le t_q) = 0 \tag{10.5}$$

M. Ochmann and R. Piscoya, *Theory and Application of Acoustic Sources Using Complex Analysis*, https://doi.org/10.1007/978-981-33-6040-2_10

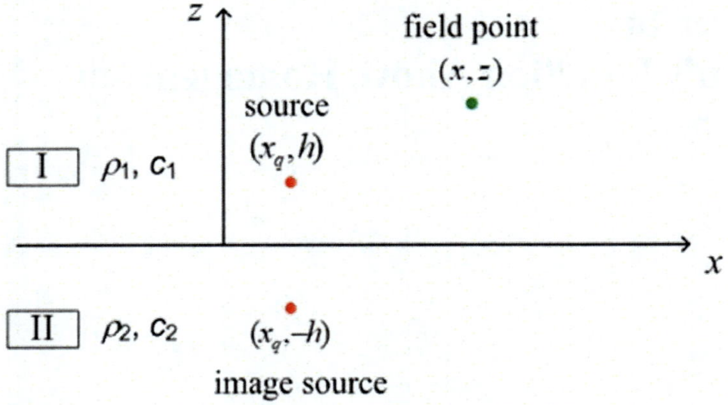

Fig. 10.1 Line source above homogeneous ground

and the convergence conditions at infinity

$$\phi_I|_{\sqrt{x^2+z^2}\to\infty} = \left.\frac{\partial\phi_I}{\partial x}\right|_{\sqrt{x^2+z^2}\to\infty} = \left.\frac{\partial\phi_I}{\partial z}\right|_{\sqrt{x^2+z^2}\to\infty} = 0, \tag{10.6}$$

$$\phi_{II}|_{\sqrt{x^2+z^2}\to\infty} = \left.\frac{\partial\phi_{II}}{\partial x}\right|_{\sqrt{x^2+z^2}\to\infty} = \left.\frac{\partial\phi_{II}}{\partial z}\right|_{\sqrt{x^2+z^2}\to\infty} = 0. \tag{10.7}$$

To solve the problem, we follow the approach from Aki and Richards [Aki09] and perform a Fourier transform in x and a Laplace transform in t. We multiply Eqs. (10.1) and (10.2) by $\mathrm{e}^{-jk_x}\mathrm{e}^{-st}$ and integrate over $\int_{-\infty}^{+\infty}\mathrm{d}k_x\int_0^{+\infty}\mathrm{d}t$.

By integration by parts and taking into account the previous conditions, Eqs. (10.1) and (10.2) become

$$n_1^2\phi_I(k_x, z, s) - \frac{\partial^2}{\partial z^2}\phi_I(k_x, z, s) = \mathrm{e}^{-jk_x x_q}\mathrm{e}^{-jst_q}\delta(z-h), \quad z > 0, \tag{10.8}$$

$$n_2^2\phi_{II}(k_x, z, s) - \frac{\partial^2}{\partial z^2}\phi_{II}(k_x, z, s) = 0, \quad z < 0, \tag{10.9}$$

with $\phi_{I,II}(k_x, z, s) = \int_{-\infty}^{+\infty}\int_0^{+\infty}\phi_{I,II}(x, z, t)\mathrm{e}^{-jk_x x}\mathrm{e}^{-jst}\mathrm{d}x\mathrm{d}t$ and $n_i = \sqrt{\left(\frac{s^2}{c_i^2}+k_x^2\right)}$.

The solution of Eq. (10.8) has the form

$$\phi_I(k_x, z, s) = \begin{cases} a\mathrm{e}^{n_1 z} + b\mathrm{e}^{-n_1 z} & 0 \le z \le h \\ c\mathrm{e}^{-n_1 z} & z \ge h \end{cases}, \tag{10.10}$$

with the boundary conditions

$$\lim_{\Delta\to 0} \phi_I\Big|_{z=h-\Delta} = \lim_{\Delta\to 0} \phi_I\Big|_{z=h+\Delta}, \tag{10.11}$$

$$\lim_{\Delta\to 0}\left(\frac{\partial\phi_I}{\partial z}\Big|_{h+\Delta} - \frac{\partial\phi_I}{\partial z}\Big|_{h-\Delta}\right) = -\mathrm{e}^{-jk_x x_q}\mathrm{e}^{-st_q} \tag{10.12}$$

and the solution of Eq. (10.9) has the form

$$\phi_{II}(k_x, z, s) = d\mathrm{e}^{-n_2 z}, \quad z < 0. \tag{10.13}$$

Applying the conditions (10.11) and (10.12), the solution in region I is given by

$$\phi_I(k_x, z, s) = \frac{\mathrm{e}^{-jk_x x_q}\mathrm{e}^{-st_q}}{2n_1}\mathrm{e}^{-n_1|z-h|} + b\mathrm{e}^{-n_1 z}. \tag{10.14}$$

The constants b and d are obtained by enforcing the boundary conditions (10.3) and (10.4). Hence, the velocity potential is given by

$$\phi_I(k_x, z, s) = \frac{\mathrm{e}^{-jk_x x_q}\mathrm{e}^{-st_q}}{2n_1}\left[\mathrm{e}^{-n_1|z-h|} + \left(\frac{\rho_2 n_1 - \rho_1 n_2}{\rho_2 n_1 + \rho_1 n_2}\right)\mathrm{e}^{-n_1(z+h)}\right], \tag{10.15}$$

$$\phi_{II}(k_x, z, s) = \frac{\mathrm{e}^{-jk_x x_q}\mathrm{e}^{-st_q}}{2n_1}\left(\frac{2\rho_1 n_1}{\rho_2 n_1 + \rho_1 n_2}\right)\mathrm{e}^{-(n_2|z|+n_1 h)}. \tag{10.16}$$

From now on, only the velocity potential in region I will be studied, therefore the subscript I will be omitted.

By applying the inverse Fourier transform on (10.15), one obtains

$$\phi(x, z, s) = \frac{1}{2\pi}\int_{-\infty}^{+\infty} \frac{\mathrm{e}^{-jk_x x_q}\mathrm{e}^{-st_q}}{2n_1}\left[\mathrm{e}^{-n_1|z-h|} + \left(\frac{\rho_2 n_1 - \rho_1 n_2}{\rho_2 n_1 + \rho_1 n_2}\right)\mathrm{e}^{-n_1(z+h)}\right]\mathrm{e}^{jk_x x}\mathrm{d}k_x. \tag{10.17}$$

Since k_x is a dummy variable of integration, one can introduce the new variable p by defining $k_x = jsp$. By using the previous definition, (10.17) can be rewritten as

$$\begin{aligned}\phi(x, z, s) = \frac{1}{2\pi}\int_{-j\infty}^{+j\infty} \frac{-j}{2\eta_1}\Big[&\mathrm{e}^{-s\left(\eta_1|z-h| + p(x-x_q) + t_q\right)} \\ &+ \left(\frac{\rho_2\eta_1 - \rho_1\eta_2}{\rho_2\eta_1 + \rho_1\eta_2}\right)\mathrm{e}^{-s\left(\eta_1(z+h) + p(x-x_q) + t_q\right)}\Big]\mathrm{d}p\end{aligned}, \tag{10.18}$$

with $\eta_i = \sqrt{\frac{1}{c_i^2} - p^2}$ and $\mathrm{Re}(\eta_i) > 0$.

The first term corresponds to the incident wave and the second term to the reflected wave.

η_i is a even function in p ($\eta_i(p) = \eta_i(-p)$) and since the values of p are purely imaginary, η_i will be always real. Equation (10.18) can be written as

$$\phi(x,z,s) = \frac{1}{2\pi}\int_0^{+j\infty} \frac{-j}{\eta_1}\left[\mathrm{e}^{-s(\eta_1|z-h|+t_q)} + \left(\frac{\rho_2\eta_1 - \rho_1\eta_2}{\rho_2\eta_1 + \rho_1\eta_2}\right)\mathrm{e}^{-s(\eta_1(z+h)+t_q)}\right] \left(\frac{e^{-sp(x-x_q)} + e^{+sp(x-x_q)}}{2}\right)\mathrm{d}p \tag{10.19}$$

$\frac{\mathrm{e}^{-sp(x-x_q)}+\mathrm{e}^{sp(x-x_q)}}{2} = \mathrm{Re}\big(\mathrm{e}^{-sp(x-x_q)}\big)$ since p is purely imaginary, hence (10.19) can be written as

$$\phi(x,z,s) = \frac{1}{2\pi}\mathrm{Im}\left(\int_0^{+j\infty}\left[\frac{\mathrm{e}^{-s(\eta_1|z-h|+p(x-x_q)+t_q)}}{\eta_1} + \left(\frac{\rho_2\eta_1 - \rho_1\eta_2}{\rho_2\eta_1 + \rho_1\eta_2}\right)\frac{\mathrm{e}^{-s(\eta_1(z+h)+p(x-x_q)+t_q)}}{\eta_1}\right]\mathrm{d}p\right) \tag{10.20}$$

The idea now is to write $\phi(x,z,s)$ in the form

$$\phi(x,z,s) = \int_0^{+\infty} \text{Integrand}\cdot \mathrm{e}^{-st}\mathrm{d}t,$$

So, that the velocity potential will be given by $\phi(x,z,t) = \text{Integrand}$. To do that, it is necessary to look for the path Γ in the complex p-plane where the argument of the exponential function is real.

First, the incident wave will be studied.

$$\phi_{\mathrm{inc}}(x,z,s) = \frac{1}{2\pi}\mathrm{Im}\left(\int_0^{j\infty}\frac{\mathrm{e}^{-s(\eta_1|z-h|+p(x-x_q)+t_q)}}{\eta_1}\mathrm{d}p\right).$$

The curve Γ is obtained by making

$$t = \eta_1|z-h| + p(x-x_q) + t_q. \tag{10.21}$$

The solution of (10.21) is given by

$$p = \frac{(x - x_q)(t - t_q) \pm |z - h|\sqrt{t_0^2 - (t - t_q)^2}}{R_0^2}, \tag{10.22}$$

with $R_0 = \sqrt{(x - x_q)^2 + (z - h)^2}$, $t_0 = R_0/c_1$.

The branch cuts that provide a positive real part of η_1 are the parts of the real p-axis ($\mathrm{Im}(p) = 0$) given by

$$\left\{\mathrm{Re}(p) < -\frac{1}{c_1}\right\} \cup \left\{\mathrm{Re}(p) > \frac{1}{c_1}\right\}. \tag{10.23}$$

The curve Γ should not cross the branch cuts.

For $(x - x_q) > 0$, by taking the negative sign in (10.22) and $\sqrt{t_0^2 - (t - t_q)^2} = -j\sqrt{(t - t_q)^2 - t_0^2}$ for $(t - t_q) > t_0$, the part of Γ on the real p-axis lies between the branch cuts ($\Gamma_0 + \Gamma_1$) and the rest of Γ (Γ_2) lies in the first quadrant where the imaginary parts of η_i are negative.

For $(x - x_q) < 0$, by taking the positive sign in (10.22) and $\sqrt{t_0^2 - (t - t_q)^2} = j\sqrt{(t - t_q)^2 - t_0^2}$ for $(t - t_q) > t_0$, the part of Γ on the real p-axis ($\Gamma_0 + \Gamma_1$) lies again between the branch cuts and the rest of Γ (Γ_2) lies in the second quadrant where the imaginary parts of η_i are positive.

The curve $\Gamma = \Gamma_0 + \Gamma_1 + \Gamma_2$ is shown in Fig. 10.2. Γ_j is the original integration path of (10.20).

At time $t - t_q = 0$, $|p| = \left(\frac{|z-h|}{R_0}\right)\frac{1}{c_1}$ and the curve Γ runs parallel to the real p-axis, passing through the origin $p = 0$ at the time $t - t_q = t_z^-$ with $t_z^- = |z - h|/c_1$, and continue until time $t - t_q = t_0$, $|p| = \left(\frac{|x-x_q|}{R_0}\right)\frac{1}{c_1}$ where p becomes complex with positive imaginary part. For large values of t, the curve Γ goes parallel to the line defined by the angle $\theta_0 = \tan^{-1}(|x - x_q|/|z - h|)$.

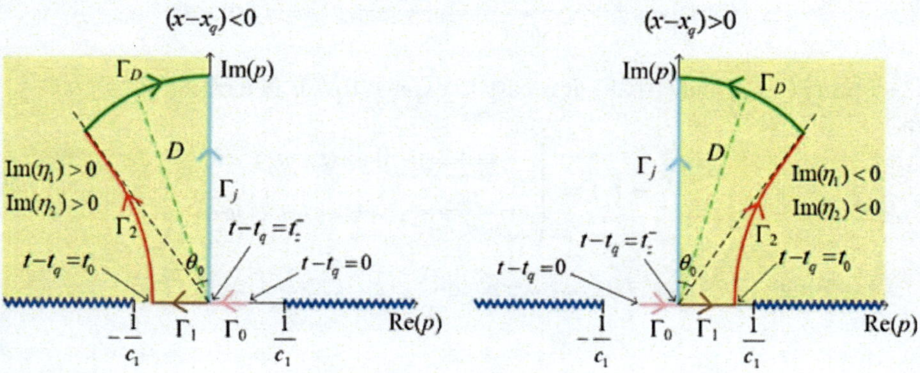

Fig. 10.2 Illustration of the alternative integration curve

To avoid the need of distinguishing between $(x - x_q) > 0$ and $(x - x_q) < 0$, it is practical to use only one curve (Fig. 10.2, right) by defining it as

$$p(t-t_q) = \begin{cases} \frac{|x-x_q|(t-t_q)-|z-h|\sqrt{t_0^2-(t-t_q)^2}}{R_0^2}, & 0 < (t-t_q) < t_0 \\ \frac{|x-x_q|(t-t_q)+j|z-h|\sqrt{(t-t_q)^2-t_0^2}}{R_0^2}, & (t-t_q) > t_0 \end{cases}, \tag{10.24}$$

which is reasonable, since the potential of a 2D monopole is expected to be symmetric with respect to a vertical axis passing through x_q.

Applying the Cauchy integral theorem over the closed path $\Gamma_1 + \Gamma_2 + \Gamma_D - \Gamma_j$ and since there are no poles inside the closed path we have

$$\lim_{D\to\infty}\left(\int_{\Gamma_1+\Gamma_2} + \int_{\Gamma_D} - \int_{\Gamma_j}\right) = 0. \tag{10.25}$$

It can be shown that $\lim_{D\to\infty}\int_{\Gamma_D} = 0$, hence

$$\int_{\Gamma_j} \frac{e^{-s\left(\eta_1(p)|z-h|+p(x-x_q)+t_q\right)}}{\eta_1(p)}\mathrm{d}p = \int_{\Gamma_1+\Gamma_2} \frac{e^{-st(p(t))}}{\eta_1(p(t))}\frac{\mathrm{d}p}{\mathrm{d}t}\mathrm{d}t. \tag{10.26}$$

We can write η_1 and $\mathrm{d}p/\mathrm{d}t$ explicitly as

$$\eta_1 = \begin{cases} \frac{|x-x_q|\sqrt{t_0^2-(t-t_q)^2}+|z-h|(t-t_q)}{R_0^2}, & 0 < (t-t_q) < t_0 \\ \frac{|z-h|(t-t_q)-j|x-x_q|\sqrt{(t-t_q)^2-t_0^2}}{R_0^2}, & (t-t_q) > t_0 \end{cases}, \tag{10.27}$$

$$\frac{\mathrm{d}p}{\mathrm{d}t}(t-t_q) = \begin{cases} \frac{|x-x_q|\sqrt{t_0^2-(t-t_q)^2}+|z-h|(t-t_q)}{R_0^2\sqrt{t_0^2-(t-t_q)^2}}, & 0 < (t-t_q) < t_0 \\ \frac{|x-x_q|\sqrt{(t-t_q)^2-t_0^2}+j|z-h|(t-t_q)}{R_0^2\sqrt{(t-t_q)^2-t_0^2}}, & (t-t_q) > t_0 \end{cases}. \tag{10.28}$$

From (10.27) and (10.28), it is easy to express $\mathrm{d}p/\mathrm{d}t$ in terms of η_1

$$\frac{\mathrm{d}p}{\mathrm{d}t}(t-t_q) = \begin{cases} \frac{\eta_1}{\sqrt{t_0^2-(t-t_q)^2}}, & 0 < (t-t_q) < t_0 \\ \frac{j\eta_1}{\sqrt{(t-t_q)^2-t_0^2}}, & (t-t_q) > t_0 \end{cases}. \tag{10.29}$$

Inserting (10.29) into (10.26), we have

$$\int_0^{j\infty} \frac{e^{-s\left(\eta_1|z-h|+p(x-x_q)+t_q\right)}}{\eta_1} dp = \int_{t_q+t_z^-}^{t_q+t_0} \frac{e^{-st}}{\sqrt{t_0^2-(t-t_q)^2}} dt + \int_{t_q+t_0}^{+\infty} \frac{je^{-st}}{\sqrt{(t-t_q)^2-t_0^2}} dt . \tag{10.30}$$

The incident wave is written as

$$\phi_{\text{inc}}(x,z,s) = \frac{1}{2\pi}\text{Im}\left(\int_{t_q+t_z^-}^{t_q+t_0} \frac{e^{-st}}{\sqrt{t_0^2-(t-t_q)^2}} dt + \int_{t_q+t_0}^{+\infty} \frac{je^{-st}}{\sqrt{(t-t_q)^2-t_0^2}} dt \right).$$

Since the first integral is purely real,

$$\phi_{\text{inc}}(x,z,s) = \frac{1}{2\pi} \int_{t_q+t_0}^{+\infty} \frac{e^{-st}}{\sqrt{(t-t_q)^2-t_0^2}} dt$$

and the incident velocity potential becomes

$$\phi_{\text{inc}}(x,z,t) = \frac{1}{2\pi} \frac{H(t-t_q-t_0)}{\sqrt{(t-t_q)^2-t_0^2}}, \tag{10.31}$$

where $H(x)$ is the unit step function.

Now, the reflected wave will be studied

$$\phi_{\text{ref}}(x,z,s) = \frac{1}{2\pi}\text{Im}\left(\int_0^{+j\infty} \left(\frac{\rho_2\eta_1-\rho_1\eta_2}{\rho_2\eta_1+\rho_1\eta_2}\right) \frac{e^{-s\left(\eta_1(z+h)+p(x-x_q)+t_q\right)}}{\eta_1} dp \right).$$

The curve Γ that makes the exponential term real is obtained by making

$$t = \eta_1(z+h) + p(x-x_q) + t_q. \tag{10.32}$$

The solution of (10.32) is given by

$$p = \frac{(x-x_q)(t-t_q) \pm (z+h)\sqrt{t_{0s}^2-(t-t_q)^2}}{R_{0s}^2}, \tag{10.33}$$

with $R_{0s} = \sqrt{(x-x_q)^2+(z+h)^2}$, $t_{0s} = R_{0s}/c_1$.

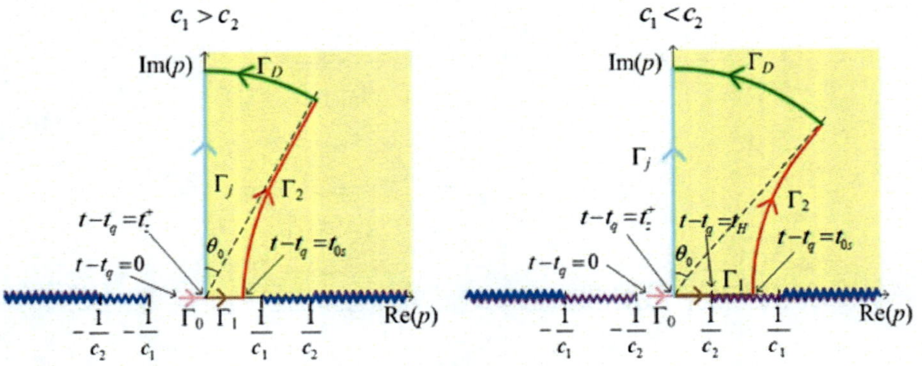

Fig. 10.3 Integration paths of the reflected wave for $c_1 > c_2$ and $c_1 < c_2$

Similarly to (10.23), the branch cuts that provide a positive real part of η_2 are the parts of the real p-axis ($\mathrm{Im}(p) = 0$) given by

$$\left\{\mathrm{Re}(p) < -\frac{1}{c_2}\right\} \cup \left\{\mathrm{Re}(p) > \frac{1}{c_2}\right\}. \tag{10.34}$$

Two possible situations arise: (a) $c_1 > c_2$ and (b) $c_1 < c_2$. In the first case, Γ_1 lies completely between the branch cuts (Fig. 10.3, left), i.e. along Γ_1, η_1 and η_2 are purely real. In the second case, part of Γ_1 lies on the branch cut, $1/c_2 < p < 1/c_1$ (Fig. 10.3, right), i.e. along that part of Γ_1, η_2 is purely imaginary and negative while η_1 remains purely real. Figure 10.3 shows the integration curves in the two cases.

Similar to the case of the incident wave, the curve Γ will be given by

$$p(t - t_q) = \begin{cases} \frac{|x - x_q|(t-t_q) - (z+h)\sqrt{t_{0s}^2 - (t-t_q)^2}}{R_{0s}^2}, & 0 < (t - t_q) < t_0 \\ \frac{|x - x_q|(t-t_q) + j(z+h)\sqrt{(t-t_q)^2 - t_{0s}^2}}{R_{0s}^2}, & (t - t_q) > t_0 \end{cases} \tag{10.35}$$

and the derivative $\mathrm{d}p/\mathrm{d}t$ satisfies Eq. (10.29). Therefore,

$$\phi_{\mathrm{ref}}(x, z, s) = \frac{1}{2\pi}\mathrm{Im}\left(\int_{t_q+t_z^+}^{t_q+t_0} \left(\frac{\rho_2\eta_1 - \rho_1\eta_2}{\rho_2\eta_1 + \rho_1\eta_2}\right) \frac{\mathrm{e}^{-st}\mathrm{d}t}{\sqrt{t_{0s}^2 - (t-t_q)^2}} + \int_{t_q+t_0}^{+\infty} \left(\frac{\rho_2\eta_1 - \rho_1\eta_2}{\rho_2\eta_1 + \rho_1\eta_2}\right) \frac{j\mathrm{e}^{-st}\mathrm{d}t}{\sqrt{(t-t_q)^2 - t_{0s}^2}} \right). \tag{10.36}$$

In the case $c_1 > c_2$, the first integral in (10.36) is purely real since η_1 and η_2 are purely real, so it gives no contribution to the reflected potential. The reflected potential will then be written as

$$\phi_{\rm ref}(x,z,s)=\frac{1}{2\pi}\int_{t_q+t_0}^{+\infty}{\rm Re}\left(\frac{\rho_2\eta_1-\rho_1\eta_2}{\rho_2\eta_1+\rho_1\eta_2}\right)\frac{{\rm e}^{-st}{\rm d}t}{\sqrt{(t-t_q)^2-t_{0s}^2}} \tag{10.37}$$

and in physical space

$$\phi_{\rm ref}(x,z,t)=\frac{1}{2\pi}{\rm Re}\left(\frac{\rho_2\eta_1-\rho_1\eta_2}{\rho_2\eta_1+\rho_1\eta_2}\right)\frac{H(t-t_q-t_{0s})}{\sqrt{(t-t_q)^2-t_{0s}^2}}. \tag{10.38}$$

Equation (10.38) can be rewritten as

$$\phi_{\rm ref}(x,z,t)=\frac{1}{2\pi}\frac{H(t-t_q-t_{0s})}{\sqrt{(t-t_q)^2-t_{0s}^2}}+\frac{1}{2\pi}{\rm Re}\left(\frac{-2\rho_1\eta_2}{\rho_2\eta_1+\rho_1\eta_2}\right)\frac{H(t-t_q-t_{0s})}{\sqrt{(t-t_q)^2-t_{0s}^2}}. \tag{10.39}$$

The first term in (10.39) would be the potential due to the image source located at $(x_q,-h)$ and the second term a correction due to the homogeneous lower half-space.

In the case $c_1<c_2$, the first integral in (10.36) gives a contribution to the reflected potential since along the part of Γ_1 given by $1/c_2<p<1/c_1$ (or $t_H<t-t_q<t_0$), η_2 is imaginary. So, the reflected potential will then be written as

$$\begin{aligned}\phi_{\rm ref}(x,z,s)=&\int_{t_q+t_H}^{t_q+t_0}{\rm Im}\left(\frac{\rho_2\eta_1-\rho_1\eta_2}{\rho_2\eta_1+\rho_1\eta_2}\right)\frac{{\rm e}^{-st}{\rm d}t}{\sqrt{t_{0s}^2-(t-t_q)^2}}\\&+\int_{t_q+t_0}^{+\infty}{\rm Re}\left(\frac{\rho_2\eta_1-\rho_1\eta_2}{\rho_2\eta_1+\rho_1\eta_2}\right)\frac{{\rm e}^{-st}{\rm d}t}{\sqrt{(t-t_q)^2-t_{0s}^2}}\end{aligned} \tag{10.40}$$

and in physical space

$$\begin{aligned}\phi_{\rm ref}(x,z,t)=&\frac{1}{2\pi}{\rm Im}\left(\frac{\rho_2\eta_1-\rho_1\eta_2}{\rho_2\eta_1+\rho_1\eta_2}\right)\frac{H(t-t_q-t_H)-H(t-t_q-t_{0s})}{\sqrt{t_{0s}^2-(t-t_q)^2}}\\&\frac{1}{2\pi}\frac{H(t-t_q-t_{0s})}{\sqrt{(t-t_q)^2-t_{0s}^2}}\\&+\frac{1}{2\pi}{\rm Re}\left(\frac{-2\rho_1\eta_2}{\rho_2\eta_1+\rho_1\eta_2}\right)\frac{H(t-t_q-t_{0s})}{\sqrt{(t-t_q)^2-t_{0s}^2}}\end{aligned} \tag{10.41}$$

The first term in (10.41) is a contribution from the so-called head waves that arrive before the contribution from the image source ("reflected wave" strictly speaking) at $t_H<t_{0s}$. According to Aki and Richards [Aki09], the reflected wave "emerges

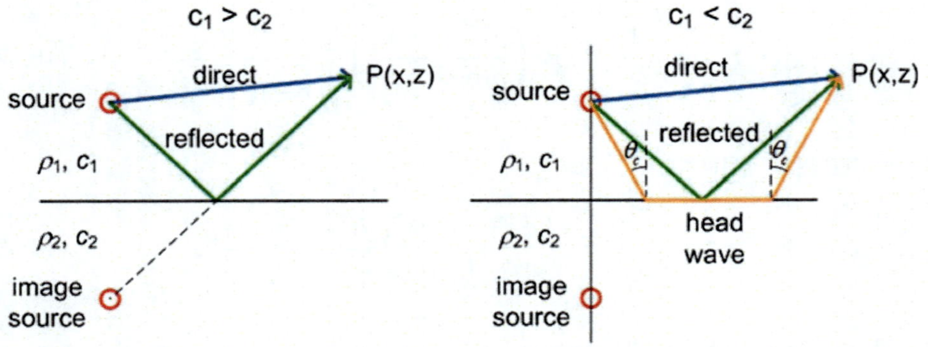

Fig. 10.4 Illustration of head waves and reflected waves

from the tail of the head wave." In terms of ray acoustics, the head wave has the path shown in Fig. 10.4, on the right, that consists on a ray hitting the ground with incident angle $\theta_{\rm i} = \theta_{\rm c}$, where $\theta_{\rm c}$ is the critical angle defined as

$$\sin\theta_{\rm c} = \frac{c_1}{c_2}, \tag{10.42}$$

followed by a ray parallel to the ground (but traveling with speed c_2) and ending with a ray coming from the ground toward the observer with reflected angle $\theta_{\rm r} = \theta_{\rm c}$.

The value of t_H is obtained by setting $p(t - t_q = t_H) = 1/c_2$ and is given by

$$t_H = \frac{|x - x_q| \sin\theta_{\rm c} + (z + h)\cos\theta_{\rm c}}{c_1}. \tag{10.43}$$

Even though (10.43) provides a value of t_H for all points in the half-space, head waves do not appear for all observer positions. They will be present at positions where $p(t - t_q = t_{0s}) > 1/c_2$, i.e., where part of Γ_1 lies on the branch cut of η_2 (Fig. 10.4)

$$p(t - t_q = t_{0s}) = \frac{|x - x_q|}{R_{0s} c_1} > \frac{1}{c_2} \to |x - x_q| > (z + h)\tan\theta_{\rm c}. \tag{10.44}$$

The same results (10.43) and (10.44) can be obtained more easily using ray acoustics. If d_1, d_2, and d_3 are the three ray paths in Fig. 10.4, on the right, the arrival time of the head waves t_H will be

$$t_H = \frac{d_1}{c_1} + \frac{d_2}{c_2} + \frac{d_3}{c_1}. \tag{10.45}$$

By replacing $d_1 = h/\cos\theta_{\rm c}$, $d_3 = z/\cos\theta_{\rm c}$ and $d_2 = |x - x_q| - d_1 \sin\theta_{\rm c} - d_3 \sin\theta_{\rm c}$ into (10.45), one obtains (10.43), and by requiring that $d_2 > 0$, one obtains (10.44).

Because part of the entire path is traveled with a speed $c_2 > c_1$, a head wave may even arrive before the direct wave. The positions where $t_H = t_0$ can be obtained analytically and are given by

$$\left|x - x_q\right| = (z + h)\tan\theta_c + \frac{2\sqrt{zh}}{\cos\theta_c}. \tag{10.46}$$

Aki and Richards [Aki09] distinguish between narrow-angle and wide-angle reflection zones:

1. narrow-angle reflection for observers at: $\frac{|x-x_q|}{R_{0s}} < \frac{c_1}{c_2}$ or $\left|x - x_q\right| < (z+h)\tan\theta_c$.
2. wide-angle reflection for observers at: $\frac{c_1}{c_2} < \frac{|x-x_q|}{R_{0s}} < 1$ or $\left|x - x_q\right| > (z + h)\tan\theta_c$.

Figure 10.5 shows the two reflection zones and the areas where $t_H < t_0$ and $t_H > t_0$.

For the case $c_1 > c_2$, the whole upper half-space belongs to the zone of narrow-angle reflection.

To summarize the results obtained up to now, the velocity potential of the line source above a homogeneous ground consists of an incident and a mirror reflected waves plus a correction term.

$$\phi(\vec{r} - \vec{r}_q; t - t_q) = \phi_{\text{inc}}(\vec{r} - \vec{r}_q; t - t_q) + \phi_{\text{mrf}}(\vec{r} - \vec{r}_q; t - t_q) + \phi_{\text{cor}}. \tag{10.47}$$

The incident and mirror reflected waves are defined as

$$\phi_{\text{inc}} = \frac{1}{2\pi}\frac{H(t - t_q - t_0)}{\sqrt{(t - t_q)^2 - t_0^2}}, \tag{10.48}$$

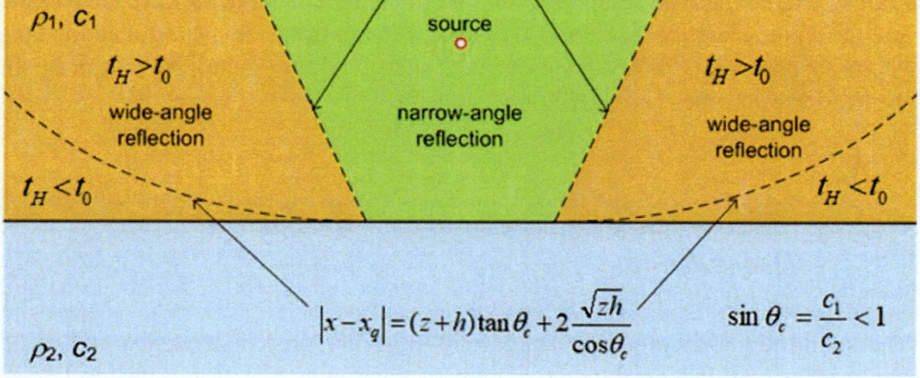

Fig. 10.5 Narrow-angle and wide-angle reflection zones

$$\phi_{\mathrm{mrf}} = \frac{1}{2\pi} \frac{H(t - t_q - t_{0s})}{\sqrt{(t - t_q)^2 - t_{0s}^2}} \tag{10.49}$$

and the correction term is given by

$$\phi_{\mathrm{cor}} = \begin{cases} \frac{1}{2\pi} \mathrm{Re}\big(\chi(\vec{r} - \vec{r}_q; t - t_q)\big) \frac{H(t-t_q-t_{0s})}{\sqrt{(t-t_q)^2 - t_{0s}^2}}, & \left|x - x_q\right| < (z+h)\tan\theta_c \\ \frac{1}{2\pi}\left[\mathrm{Im}\big(\chi(\vec{r} - \vec{r}_q; t - t_q)\big) \frac{H(t - t_q - t_H) - H(t - t_q - t_{0s})}{\sqrt{t_{0s}^2 - (t - t_q)^2}} \right. & \\ \left. + \mathrm{Re}\big(\chi(\vec{r} - \vec{r}_q; t - t_q)\big) \frac{H(t - t_q - t_{0s})}{\sqrt{(t - t_q)^2 - t_{0s}^2}} \right], & \left|x - x_q\right| > (z+h)\tan\theta_c \end{cases} \tag{10.50}$$

where χ is the coefficient

$$\chi = -\frac{2\rho_1\eta_2(\vec{r} - \vec{r}_q; t - t_q)}{\rho_2\eta_1(\vec{r} - \vec{r}_q; t - t_q) + \rho_1\eta_2(\vec{r} - \vec{r}_q; t - t_q)}, \tag{10.51}$$

with

$$\eta_i = \sqrt{\frac{1}{c_i^2} - p^2(\vec{r} - \vec{r}_q; t - t_q)}, \tag{10.52}$$

$$p = \begin{cases} \frac{\left|x - x_q\right|(t - t_q) - (z+h)\sqrt{t_{0s}^2 - (t - t_q)^2}}{R_{0s}^2} & t - t_q < t_{0s} \\ \frac{\left|x - x_q\right|(t - t_q) + j(z+h)\sqrt{(t - t_q)^2 - t_{0s}^2}}{R_{0s}^2} & t - t_q > t_{0s} \end{cases}. \tag{10.53}$$

Figure 10.6 shows the incident and reflected waves at three different positions and $(x_q, h) = (0, 2.5)$. The parameters of the second medium are $c_2 = 2.5c_1$ and $\rho_2 = 4\rho_1$. For those values, the critical angle is $\theta_c = 23.6°$. The first point is located at (1.5, 1.5) in the narrow-angle reflection zone, so there is no head wave. The second and third points are located at (4, 1) and (4, 0.25) in the wide-angle reflection zone. At the second point, the head wave arrives after the direct sound. At the third point, the head wave arrives before the direct sound.

10.1.2 Sound Field of a Line Source with Arbitrary Time Dependence

The sound radiation of a line source with arbitrary time dependence $q(t)$ can be obtained by performing the convolution of the source function with the Green's function

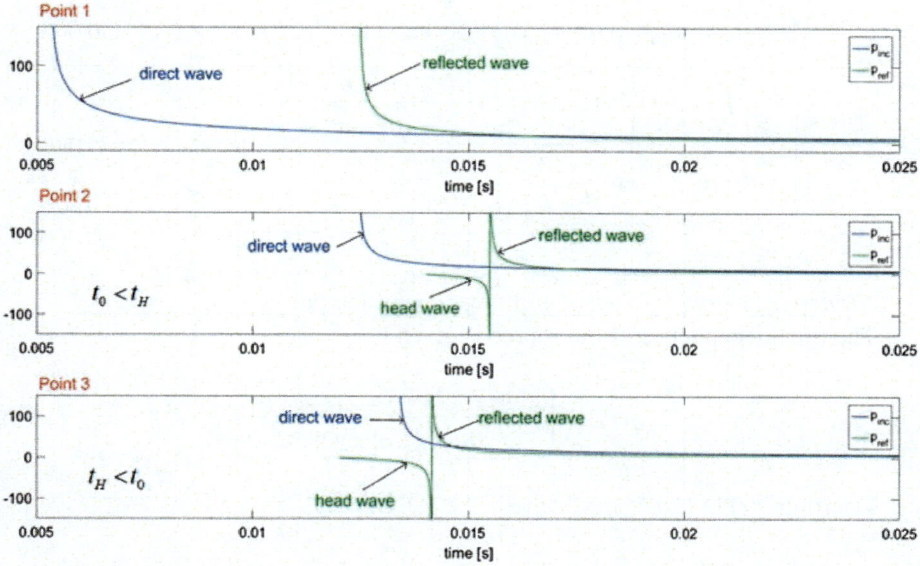

Fig. 10.6 Incident and reflected wave at three different positions

$$\Phi(\vec{r}, t) = \int_{-\infty}^{+\infty} q(\tau)\phi(\vec{r} - \vec{r}_q; t - \tau)\mathrm{d}\tau. \tag{10.54}$$

The incident and reflected waves are given by

$$\begin{aligned}\Phi_{\text{inc}} &= \frac{1}{2\pi}\int_{-\infty}^{+\infty} q(\tau)\frac{H(t-\tau-t_0)}{\sqrt{(t-\tau)^2 - t_0^2}}\mathrm{d}\tau \\ &= \frac{1}{2\pi}\int_{-\infty}^{+\infty} q(t-\tau)\frac{H(\tau-t_0)}{\sqrt{\tau^2 - t_0^2}}\mathrm{d}\tau,\end{aligned} \tag{10.55}$$

$$\begin{aligned}\Phi_{\text{mrf}} &= \frac{1}{2\pi}\int_{-\infty}^{+\infty} q(\tau)\frac{H(t-\tau-t_{0s})}{\sqrt{(t-\tau)^2 - t_{0s}^2}}\mathrm{d}\tau \\ &= \frac{1}{2\pi}\int_{-\infty}^{+\infty} q(t-\tau)\frac{H(\tau-t_{0s})}{\sqrt{\tau^2 - t_{0s}^2}}\mathrm{d}\tau\end{aligned} \tag{10.56}$$

and the correction term will be expressed as

$$\Phi_{\text{cor}} = \begin{cases} \frac{1}{2\pi}\int\limits_{-\infty}^{+\infty} q(t-\tau)\mathrm{Re}\big(\chi(\vec{r}-\vec{r}_q;\tau)\big)\frac{H(\tau-t_{0s})}{\sqrt{\tau^2-t_{0s}^2}}\mathrm{d}\tau, & |x-x_q| < (z+h)\tan\theta_c \\ \frac{1}{2\pi}\int\limits_{-\infty}^{+\infty} q(t-\tau)\Bigg[\mathrm{Im}\big(\chi(\vec{r}-\vec{r}_q;\tau)\big)\frac{H(\tau-t_H)-H(\tau-t_{0s})}{\sqrt{t_{0s}^2-\tau^2}} & \\ \quad +\mathrm{Re}\big(\chi(\vec{r}-\vec{r}_q;\tau)\big)\frac{H(\tau-t_{0s})}{\sqrt{\tau^2-t_{0s}^2}}\Bigg]\mathrm{d}\tau, & |x-x_q| > (z+h)\tan\theta_c \end{cases} \tag{10.57}$$

Numerical example: Source emitting a Gaussian pulse
The time dependence of the source is given by

$$q(t) = \begin{cases} \mathrm{e}^{-(t-\mu)^2/2\sigma^2} & |t-\mu| < \Delta \\ 0 & \text{elsewhere} \end{cases}. \tag{10.58}$$

Inserting $q(t)$ in (10.55), we have

$$\phi_{\text{inc}}(t) = \frac{1}{2\pi}\int\limits_{t_0}^{\infty} \frac{\mathrm{e}^{-(t-\mu-\tau)^2/2\sigma^2}}{\sqrt{\tau^2-t_0^2}}(H(t-\mu+\Delta-\tau) - H(t-\mu-\Delta-\tau))\mathrm{d}\tau \tag{10.59}$$

The integration range is the intersection of two intervals, a semi-infinite fixed one $[t_0, \infty[$ and a finite moving one $[t-\mu-\Delta, t-\mu+\Delta]$ (Fig. 10.7).

The incident wave can be written as

$$\phi_{\text{inc}}(t) = \frac{1}{2\pi}\int\limits_{\tau_1}^{\tau_2} \frac{\mathrm{e}^{-(t-\mu-\tau)^2/2\sigma^2}}{\sqrt{\tau^2-t_0^2}}\mathrm{d}\tau \tag{10.60}$$

with $\tau_1 = \max(t-\mu-\Delta, t_0)$ and $\tau_2 = \max(t-\mu+\Delta, t_0)$ for $\tau_2 > \tau_1$.

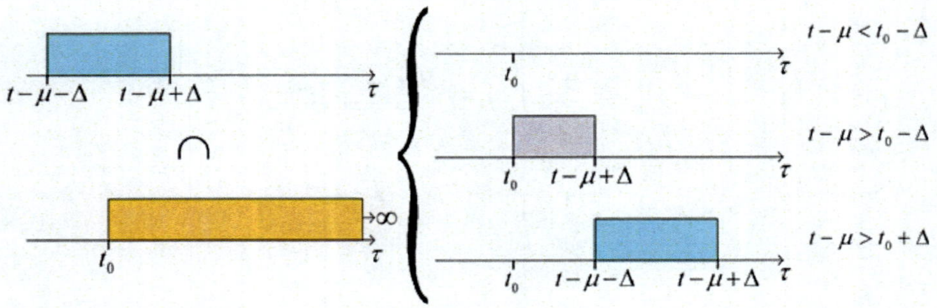

Fig. 10.7 Illustration of the integration range

To avoid the singularity in (10.60), we use the change of variable $\tau = t_0 \cosh u$ and obtain

$$\phi_{\text{inc}}(t) = \frac{1}{2\pi} \int_{u_1}^{u_2} \mathrm{e}^{-(t-\mu-t_0 \cosh u)^2/2\sigma^2} \mathrm{d}u, \tag{10.61}$$

with $u_1 = \operatorname{a}\cosh(\tau_1/t_0)$ and $u_2 = \operatorname{a}\cosh(\tau_2/t_0)$.

In a similar way, the mirror reflected wave can be obtained from (10.56) and is given by

$$\phi_{\text{mrf}}(t) = \frac{1}{2\pi} \int_{u_{1s}}^{u_{2s}} \mathrm{e}^{-(t-\mu-t_{0s} \cosh u)^2/2\sigma^2} \mathrm{d}u, \tag{10.62}$$

with $u_{1s} = \operatorname{a}\cosh(\tau_{1s}/t_{0s})$ and $u_{2s} = \operatorname{a}\cosh(\tau_{2s}/t_{0s})$, where $\tau_{1s} = \max(t-\mu-\Delta, t_{0s})$ and $\tau_{2s} = \max(t-\mu+\Delta, t_{0s})$ for $\tau_{2s} > \tau_{1s}$.

The correction term (10.57) has two integrals. The integration range is finite in both integrals since the source term is also finite. The first integral I_1 follows the same analysis as for the mirror reflected wave, so it is written as

$$I_1(t) = \frac{1}{2\pi} \int_{u_{1s}}^{u_{2s}} \mathrm{e}^{-(t-\mu-t_{0s} \cosh u)^2/2\sigma^2} \mathrm{Re}(\Gamma(x, z, t_{0s} \cosh u)) \mathrm{d}u. \tag{10.63}$$

For the second integral I_2, the integration range is the intersection of two finite intervals, a fixed one $[t_H, t_{0s}]$ and a moving one $[t-\mu-\Delta, t-\mu+\Delta]$ (Fig. 10.8).

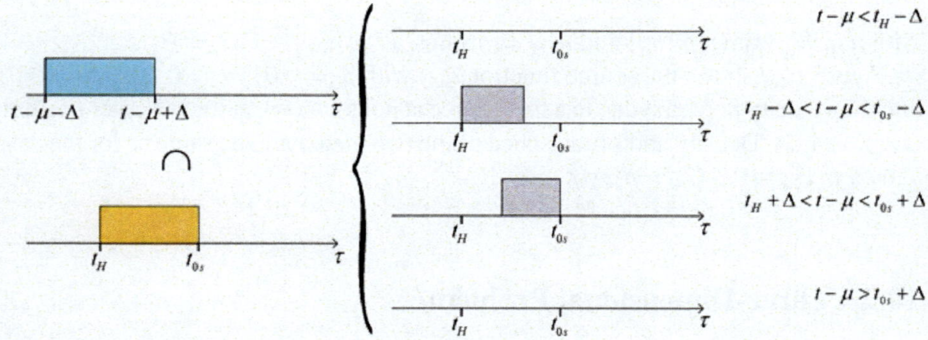

Fig. 10.8 Illustration of the integration range of I_2

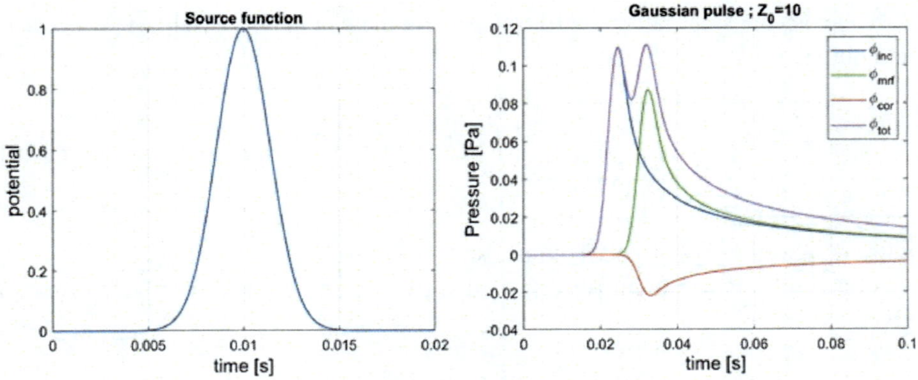

Fig. 10.9 Gaussian pulse (left) and potential at the observer position (right)

The integral considering the contribution of the head waves reads

$$I_2(t) = \frac{1}{2\pi} \int_{\tau_{1H}}^{\tau_{2H}} \frac{e^{-(t-\mu-\tau)^2/2\sigma^2}}{\sqrt{t_{0s}^2 - \tau^2}} \mathrm{Im}(\Gamma(x, z, \tau))\mathrm{d}\tau, \tag{10.64}$$

with $\tau_{1H} = \min(\max(t - \mu - \Delta, t_H), t_{0s})$ and $\tau_{2H} = \min(\max(t - \mu + \Delta, t_H), t_{0s})$ for $\tau_{2H} > \tau_{1H}$.

To avoid the singularity in (10.64), we use the change of variable $\tau = t_{0s} \sin u$ and obtain

$$I_2(t) = \frac{1}{2\pi} \int_{u_{1H}}^{u_{2H}} e^{-(t-\mu-t_0 \sin u)^2/2\sigma^2} \mathrm{Im}(\Gamma(x, z, t_{0s} \sin u))\mathrm{d}u, \tag{10.65}$$

with $u_{1H} = a\sin(\tau_{1H}/t_{0s})$ and $u_{2H} = a\sin(\tau_{2H}/t_{0s})$.

Figure 10.9 shows the source function $q(t)$ with $\mu = 0.01$, $\sigma = 0.002$, $\Delta = 0.01$ and the velocity potential due to a line source at height $h = 4$, at the observer position $(x, z) = (4, 2)$. Density and sound speed of the two media are the same as for the case shown in Fig. 10.6 ($Z_0 = \rho_2 c_2/\rho_1 c_1 = 10$).

10.2 Three-Dimensional Problem

In this section, we will show two different approaches to determine the velocity potential due to a point source with harmonic time dependence above the homogeneous ground, the Sommerfeld approach which uses an expansion in cylindrical waves, and the exact image theory (EIT) developed by Lindell [Lin86] that uses an expansion in spherical waves. The results of both approaches are compared.

10.2.1 Sommerfeld Approach

For the time convention $e^{j\omega t}$, the velocity potential is assumed to be of the form

$$\phi = \begin{cases} Q_0 \frac{e^{-jk_1 R_1}}{4\pi R_1} + Q_0 \frac{e^{-jk_1 R_2}}{4\pi R_2} + C_1 & z > 0 \\ C_2 & z < 0 \end{cases}, \tag{10.66}$$

where Q_0 is the volume velocity of the source and R_1 and R_2 are the distances from the source and the image source to the field point, respectively.

By considering the Sommerfeld integral identity, the first two terms can be defined as

$$\frac{e^{-jk_1 R_1}}{4\pi R_1} = \frac{1}{4\pi} \int_0^\infty J_0(\lambda\rho) e^{-\ell|z-h|} \frac{\lambda}{\ell} d\lambda, \tag{10.67}$$

$$\frac{e^{-jk_1 R_2}}{4\pi R_2} = \frac{1}{4\pi} \int_0^\infty J_0(\lambda\rho) e^{-\ell(z+h)} \frac{\lambda}{\ell} d\lambda, \tag{10.68}$$

where $\ell = \sqrt{\lambda^2 - k_1^2}$. This identity shows that a spherical wave can be expanded into a sum of cylindrical waves.

The idea is to define C_1 and C_2 in a similar way as (10.67) and (10.68)

$$C_1 = \frac{1}{4\pi} \int_0^\infty J_0(\lambda\rho) e^{-\ell(z+h)} f_1(\lambda) \frac{\lambda}{\ell} d\lambda, \tag{10.69}$$

$$C_2 = \frac{1}{4\pi} \int_0^\infty J_0(\lambda\rho) e^{mz} f_2(\lambda) \frac{\lambda}{m} d\lambda, \tag{10.70}$$

where $m = \sqrt{\lambda^2 - k_2^2}$ and f_1 and f_2 are two functions that will be determined by satisfying the boundary conditions on the interface

$$\rho_1 \phi_1|_{z=0} = \rho_2 \phi_2|_{z=0}, \tag{10.71}$$

$$\left.\frac{\partial \phi_1}{\partial z}\right|_{z=0} = \left.\frac{\partial \phi_2}{\partial z}\right|_{z=0}. \tag{10.72}$$

By inserting (10.67)–(10.70) into (10.71) and (10.72), one obtain the functions f_1 and f_2

$$f_1 = -\frac{2Q_0\rho_1 m}{\rho_1 m + \rho_2 \ell}, \tag{10.73}$$

$$f_2 = \frac{2Q_0\rho_1 m \mathrm{e}^{-\ell h}}{\rho_1 m + \rho_2 \ell}. \tag{10.74}$$

Therefore, the functions C_1 and C_2 are given by

$$C_1 = -\frac{\rho_1 Q_0}{2\pi} \int_0^\infty \frac{J_0(\lambda\rho)\mathrm{e}^{-\ell(z+h)}}{\rho_1 m + \rho_2 \ell} \frac{m}{\ell} \lambda \mathrm{d}\lambda, \tag{10.75}$$

$$C_2 = \frac{\rho_1 Q_0}{2\pi} \int_0^\infty \frac{J_0(\lambda\rho)\mathrm{e}^{-\ell h}\mathrm{e}^{mz}}{\rho_1 m + \rho_2 \ell} \mathrm{d}\lambda. \tag{10.76}$$

Since medium 1 is assumed to be air, the wave number k_1 is purely real while k_2 can be in general a complex number. In order to have convergent integrals, the real parts of ℓ and m must be positive and to satisfy the radiation condition, the imaginary part of ℓ and m must be also positive

$$\mathrm{Re}\left(\sqrt{\lambda^2 - k_1^2}\right) > 0, \mathrm{Im}\left(\sqrt{\lambda^2 - k_1^2}\right) > 0, \tag{10.77}$$

$$\mathrm{Re}\left(\sqrt{\lambda^2 - k_2^2}\right) > 0, \mathrm{Im}\left(\sqrt{\lambda^2 - k_2^2}\right) > 0. \tag{10.78}$$

The integral (10.75) has a singularity at the point $\lambda = k_1$. This singularity is integrable and can be numerically overcome by going around the singularity point into the complex plane (Fig. 10.10).

10.2.2 Exact Image Theory (EIT)

Here, the time convention $\mathrm{e}^{-j\omega t}$ is used. The wave equations to be solved in both regions are given by

$$\Delta\phi_1 + k_1^2\phi_1 = -Q_0\delta(x - x_q)\delta(y - y_q)\delta(z - h), \tag{10.79}$$

$$\Delta\phi_2 + k_2^2\phi_2 = 0. \tag{10.80}$$

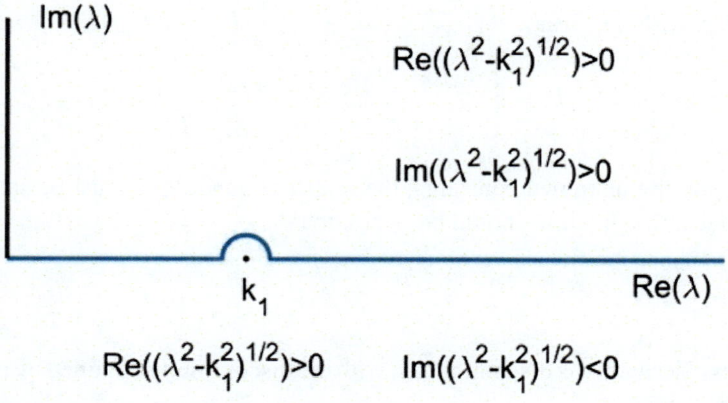

Fig. 10.10 Integration path for the integral (10.75)

Using the Fourier transform, the velocity potentials can be written as

$$\phi_i(x, y, z) = \frac{1}{(2\pi)^2} \int_{-\infty}^{+\infty} \int_{-\infty}^{+\infty} \hat{\phi}_i(k_x, k_y, z) \mathrm{e}^{jk_x(x-x_q)} \mathrm{e}^{jk_y(y-y_q)} \mathrm{d}k_x \mathrm{d}k_y, \quad i = 1, 2. \tag{10.81}$$

By definition, the delta functions are given by

$$\delta(x - x_q) = \frac{1}{2\pi} \int_{-\infty}^{+\infty} \mathrm{e}^{jk_x(x-x_q)} \mathrm{d}k_x,$$
$$\delta(y - y_q) = \frac{1}{2\pi} \int_{-\infty}^{+\infty} \mathrm{e}^{jk_y(y-y_q)} \mathrm{d}k_y. \tag{10.82}$$

Inserting (10.81) and (10.82) into (10.79), we obtain

$$\frac{1}{(2\pi)^2} \int_{-\infty}^{+\infty} \int_{-\infty}^{+\infty} \left[\left(k_1^2 - k_x^2 - k_y^2 + \frac{\mathrm{d}^2}{\mathrm{d}z^2} \right) \hat{\phi}_1(k_x, k_y, z) + Q_0 \delta(z - h) \right]$$
$$\mathrm{e}^{jk_x(x-x_q)} \mathrm{e}^{jk_y(y-y_q)} \mathrm{d}k_x \mathrm{d}k_y = 0. \tag{10.83}$$

The function $\hat{\phi}_1$ must satisfy the equation

$$\frac{\mathrm{d}^2 \hat{\phi}_1}{\mathrm{d}z^2} + k_{1z}^2 \hat{\phi}_1 = -Q_0 \delta(z - h), \; k_{1z} = \sqrt{k_1^2 - k_x^2 - k_y^2}. \tag{10.84}$$

We start with the ansatz

$$\hat{\phi}_1 = \begin{cases} Ae^{jk_{1z}z} + Be^{-jk_{1z}z} & 0 < z < h \\ Ce^{jk_{1z}z} & z > h \end{cases}. \tag{10.85}$$

To obtain the unknown constants, the interface conditions must be satisfied. On the interface, the function should be continuous

$$\hat{\phi}_1\Big|_{z=h^+} = \hat{\phi}_1\Big|_{z=h^-} \tag{10.86}$$

and the first derivative is discontinuous, with the discontinuity obtained by integrating Eq. (10.84)

$$\lim_{\varepsilon\to 0}\int_{h-\varepsilon}^{h+\varepsilon}\left(\frac{d^2\hat{\phi}_1}{dz^2} + k_{1z}^2\hat{\phi}_1\right)dz = -Q_0\lim_{\varepsilon\to 0}\int_{h-\varepsilon}^{h+\varepsilon}\delta(z-h)dz,$$

$$\to \frac{d\hat{\phi}_1}{dz}\Big|_{z=h^+} - \frac{d\hat{\phi}_1}{dz}\Big|_{z=h^-} = -Q_0. \tag{10.87}$$

By inserting (10.85) into (10.86) and (10.87), we obtain

$$B = \frac{jQ_0}{2k_{1z}} \text{ and } C = A + \frac{jQ_0}{2k_{1z}}e^{-jk_{1z}h} \tag{10.88}$$

and $\hat{\phi}_1$ can be written as

$$\hat{\phi}_1 = Ae^{jk_{1z}z} + \frac{jQ_0}{2k_{1z}}e^{jk_{1z}|z-h|}. \tag{10.89}$$

The function $\hat{\phi}_2$ must satisfy the equation

$$\frac{d^2\hat{\phi}_2}{dz^2} + k_{2z}^2\hat{\phi}_2 = 0,\ k_{2z} = \sqrt{k_2^2 - k_x^2 - k_y^2}. \tag{10.90}$$

The ansatz is

$$\hat{\phi}_2 = De^{-jk_{2z}z} \tag{10.91}$$

and the boundary conditions on the ground require continuity of pressure and normal particle velocity

$$\rho_1\hat{\phi}_1\Big|_{z=0} = \rho_2\hat{\phi}_2\Big|_{z=0}, \tag{10.92}$$

$$\left.\frac{d\hat{\phi}_1}{dz}\right|_{z=0} = \left.\frac{d\hat{\phi}_2}{dz}\right|_{z=0}. \tag{10.93}$$

By solving (10.92) and (10.93), one finds

$$A = \frac{jQ_0}{2k_{1z}}\left(\frac{\rho_2 k_{1z} - \rho_1 k_{2z}}{\rho_2 k_{1z} + \rho_1 k_{2z}}\right) \text{ and } D = \frac{jQ_0}{2k_{2z}}\left(\frac{2\rho_1 k_{2z} k_{1z}}{\rho_2 k_{1z} + \rho_1 k_{2z}} \frac{e^{jk_{1z}h}}{k_{1z}}\right), \tag{10.94}$$

where the expressions in brackets are the reflection and transmission coefficients $\hat{R}_w$ and $\hat{T}_w$, respectively.

Finally, the velocity potentials in Fourier space are written as

$$\hat{\phi}_1 = \frac{jQ_0}{2k_{1z}} e^{jk_{1z}|z-h|} + \frac{jQ_0}{2k_{1z}}\left(\frac{\rho_2 k_{1z} - \rho_1 k_{2z}}{\rho_2 k_{1z} + \rho_1 k_{2z}}\right) e^{jk_{1z}(z+h)}, \tag{10.95}$$

$$\hat{\phi}_2 = \frac{jQ_0}{2k_{2z}}\left(\frac{2\rho_1 k_{2z} k_{1z}}{\rho_2 k_{1z} + \rho_1 k_{2z}} \frac{e^{jk_{1z}h}}{k_{1z}}\right) e^{-jk_{2z}z}. \tag{10.96}$$

The first term in (10.95) is the incident wave and the second term the reflected wave and (10.96) is the transmitted wave.

In the physical space, the incident wave is given by

$$\phi_{\text{inc}} = \frac{1}{(2\pi)^2} \int_{-\infty}^{+\infty}\int_{-\infty}^{+\infty} \frac{jQ_0}{2k_{1z}} e^{jk_{1z}|z-h|} e^{jk_x(x-x_q)} e^{jk_y(y-y_q)} dk_x dk_y, \tag{10.97}$$

$$\begin{aligned} \phi_{\text{inc}} &= \frac{Q_0}{4\pi}\left(\frac{j}{2\pi}\int_{-\infty}^{+\infty}\int_{-\infty}^{+\infty} \frac{e^{jk_{1z}|z-h|} e^{jk_x(x-x_q)} e^{jk_y(y-y_q)}}{k_{1z}} dk_x dk_y\right) \\ &= \frac{Q_0}{4\pi}\frac{e^{jk_1 R_1}}{R_1} \end{aligned}. \tag{10.98}$$

To find a formula for the reflected wave, Lindell expresses the reflection coefficient in terms of k_{1z} alone by rewriting k_{2z} as

$$k_{2z} = \sqrt{B^2 + k_{1z}^2},\ B = \sqrt{k_2^2 - k_1^2} \tag{10.99}$$

and proposes to define it as the Laplace transform of a function $s(q)$

$$\hat{R}_w(k_{1z}) = \int_0^{+\infty} s_R(q) e^{-k_{1z} H_R(q)} dq. \tag{10.100}$$

Then, the reflected wave can be defined as

$$\phi_{\text{ref}} = \frac{1}{(2\pi)^2}\int_{-\infty}^{+\infty}\int_{-\infty}^{+\infty}\frac{jQ_0}{2k_{1z}}\left(\int_0^{+\infty} s_R(q)\mathrm{e}^{-k_{1z}H_R(q)}\mathrm{d}q\right)$$
$$\mathrm{e}^{jk_x(x-x_q)}\mathrm{e}^{jk_y(y-y_q)}\mathrm{e}^{jk_{1z}(z+h)}\mathrm{d}k_x\mathrm{d}k_y,$$

$$\phi_{\text{ref}} = \int_0^{+\infty} s_R(q)\frac{Q_0}{4\pi}\left(\frac{j}{2\pi}\int_{-\infty}^{+\infty}\int_{-\infty}^{+\infty}\frac{\mathrm{e}^{jk_x(x-x_q)}\mathrm{e}^{jk_y(y-y_q)}\mathrm{e}^{jk_{1z}(z+h+jH_R(q))}}{k_{1z}}\mathrm{d}k_x\mathrm{d}k_y\right)\mathrm{d}q,$$

$$\phi_{\text{ref}} = \int_0^{+\infty} s_R(q)\frac{Q_0}{4\pi}\frac{\mathrm{e}^{jk_1R_2^c(q)}}{R_2^c(q)}\mathrm{d}q, \tag{10.101}$$

where $R_2^c(q) = \sqrt{(x-x_q)^2+(y-y_q)^2+(z+h+jH_R(q))^2}$. To ensure that the integral converges and ϕ_{ref} satisfies the radiation condition, we require that real and imaginary parts of R_2^c be positive

$$\mathrm{Re}(R_2^c) > 0 \text{ and } \mathrm{Im}(R_2^c) > 0. \tag{10.102}$$

Equation (10.101) indicates that ϕ_{ref} can be represented by a distribution of point sources with density $s_R(q)$ located at complex positions $z_q(q) = -h - jH_R(q)$.

To find the source distribution $s_R(q)$, we analyze $\hat{R}_w(k_{1z})$

$$\hat{R}_w(k_{1z}) = \frac{\rho_2k_{1z} - \rho_1\sqrt{k_{1z}^2+B^2}}{\rho_2k_{1z} + \rho_1\sqrt{k_{1z}^2+B^2}}.$$

By denoting $\alpha = \frac{\rho_2}{\rho_1}$ and $p = \frac{k_{1z}}{B}$ we can write

$$\hat{R}_w(p) = \frac{\alpha p - \sqrt{p^2+1}}{\alpha p + \sqrt{p^2+1}}, \tag{10.103}$$

which is not the Laplace transform of a known function. The next step is to rearrange (10.103) in a way that the formulas from Laplace transforms of known functions can be associated.

It can be shown that

$$\hat{R}_w(p) = \frac{\alpha-1}{\alpha+1} - \alpha\gamma\left(\frac{\alpha-1}{\alpha+1}\right)\left(\frac{\gamma}{p^2-\gamma^2}\right) + \alpha\gamma\left(\frac{\gamma}{p^2-\gamma^2}\right)\left(p-\sqrt{p^2+1}\right)^2, \tag{10.104}$$

with $\gamma^2 = 1/(\alpha^2-1)$. The deduction can be found in Appendix D.

From tables of Laplace transform, one can identify the pair of functions (f, F) defined as

$$F(p) = \int_0^{+\infty} f(q)\mathrm{e}^{-qp}\mathrm{d}q. \tag{10.105}$$

Looking for each term in (10.104), one finds

$$F_o(p) = 1 \leftrightarrow f_0(q) = \delta_+(q) \quad \text{where} \int_0^{+\infty} \delta_+(q)\mathrm{d}q = 1, \tag{10.106}$$

$$F_1(p) = \frac{\gamma}{p^2 - \gamma^2} \leftrightarrow f_1(q) = \sinh(\gamma q), \tag{10.107}$$

$$F_2(p) = \left(p - \sqrt{p^2+1}\right)^2 \leftrightarrow f_2(q) = \frac{2J_2(p)}{p}, \tag{10.108}$$

$$F_1(p)F_2(p) \leftrightarrow \int_0^p f_1(p-q)f_2(q)\mathrm{d}q = \int_0^p f_1(q)f_2(p-q)\mathrm{d}q. \tag{10.109}$$

Inserting (10.105)–(10.109) into (10.104), we have

$$\begin{aligned}\hat{R}_w(p) = \int_0^{+\infty} \Bigg[&\frac{\alpha-1}{\alpha+1}\delta_+(q) - \alpha\gamma\left(\frac{\alpha-1}{\alpha+1}\right)\sinh(\gamma q) \\ &+\alpha\gamma\int_0^q \sinh(\gamma(q-u))\frac{2J_2(u)}{u}\mathrm{d}u\Bigg]\mathrm{e}^{-qp}\mathrm{d}q\end{aligned} \tag{10.110}$$

Comparing (10.110) with (10.100), we find that the source density $s_R(q)$ is

$$\begin{aligned}s_R(q) = &\frac{\alpha-1}{\alpha+1}\delta_+(q) - \alpha\gamma\left(\frac{\alpha-1}{\alpha+1}\right)\sinh(\gamma q) \\ &+\alpha\gamma\int_0^q \sinh(\gamma(q-u))\frac{2J_2(u)}{u}\mathrm{d}u\end{aligned} \tag{10.111}$$

and $H_R(q) = \frac{q}{B}$.

Inserting (10.111) into (10.101), the reflected wave is written as

$$\phi_{\text{ref}} = \left(\frac{\alpha-1}{\alpha+1}\right)\frac{Q_0 e^{jk_1 R_2}}{4\pi R_2} + \int_0^{+\infty} r(q)\frac{Q_0}{4\pi}\frac{e^{jk_1 R_2^c(q)}}{R_2^c(q)} dq, \tag{10.112}$$

with

$$r(q) = -\alpha\gamma\left(\frac{\alpha-1}{\alpha+1}\right)\sinh(\gamma q) + \alpha\gamma\int_0^q \sinh(\gamma(q-u))\frac{2J_2(u)}{u}du.$$

In order for $r(q)$ to be a valid source distribution, the source positions shall not lie in the upper half-space. For the time convention used in this section, real and imaginary parts of k_2 are positive

$$\text{Re}(k_2) > 0 \text{ and } \text{Im}(k_2) > 0 \tag{10.113}$$

and since k_1 is purely real, real and imaginary parts of B are also positive

$$\text{Re}(B) > 0 \text{ and } \text{Im}(B) > 0. \tag{10.114}$$

Therefore, we see that the source positions are given by

$$z_q(q) = -h - q\frac{\text{Im}(B)}{|B|} - jq\frac{\text{Re}(B)}{|B|} \rightarrow \text{Re}(z_q) < 0 \text{ and } \text{Im}(z_q) < 0. \tag{10.115}$$

(see Fig. 10.11). From (10.114) we verify that by choosing the branch $\text{Re}(R_2^c) > 0$, we have $\text{Im}(R_2^c) > 0$ for $z > -h - q\text{Im}(B)/|B|$, i.e., on the upper half-space ($z > 0$), the condition (10.102) is satisfied and $s_R(q)$ is a valid source distribution.

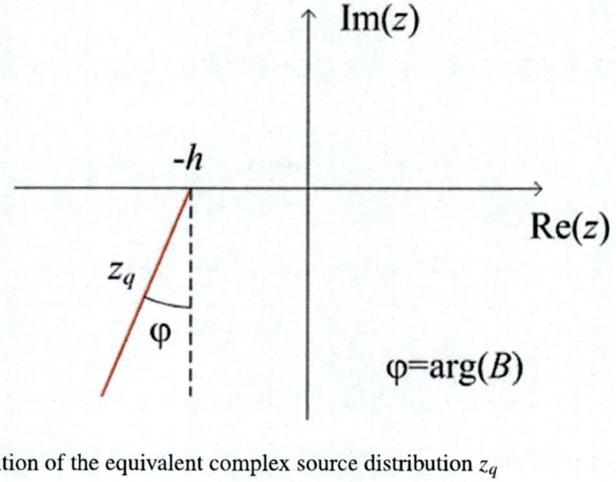

Fig. 10.11 Position of the equivalent complex source distribution z_q

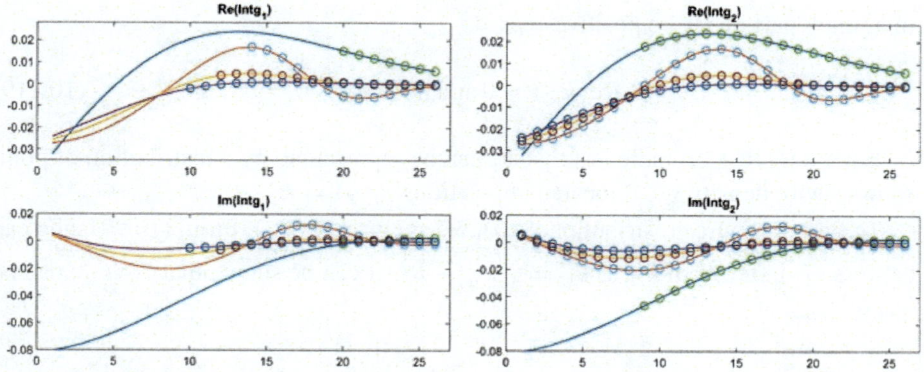

Fig. 10.12 Comparison of the results of relations (10.122) and (10.132)

For practical applications, $r(q)$ is numerically unstable due to the sinh terms. A more suitable expression for computational calculations is given by

$$r(q) = -\alpha\gamma \int_0^{+\infty} e^{-\gamma u}\left(\frac{J_2(q+u)}{q+u} + \frac{J_2(q-u)}{q-u}\right) du. \quad (10.116)$$

The deduction of (10.116) is given in Appendix D.

The last step is to find the transmitted wave. Lindell proceeds in a similar way as for the reflected wave. The idea is to represent the transmission coefficient as

$$\hat{T}_w(k_{2z}) = \int_0^{+\infty} s_T(q) e^{jk_{2z}H_T(q)} dq. \quad (10.117)$$

Then, the transmitted wave can be defined as

$$\phi_{tr} = \frac{1}{(2\pi)^2} \int_{-\infty}^{+\infty}\int_{-\infty}^{+\infty} \frac{jQ_0}{2k_{2z}} \left(\int_0^{+\infty} s_T(q) e^{jk_{2z}H_T(q)} dq\right) e^{jk_x(x-x_q)} e^{jk_y(y-y_q)} e^{-jk_{2z}z} dk_x dk_y,$$

$$\phi_{tr} = \int_0^{+\infty} s_T(q) \frac{Q_0}{4\pi} \left(\frac{j}{2\pi} \int_{-\infty}^{+\infty}\int_{-\infty}^{+\infty} \frac{e^{jk_x(x-x_q)} e^{jk_y(y-y_q)} e^{jk_{2z}(H_T(q)-z)}}{k_{2z}} dk_x dk_y\right) dq,$$

$$\phi_{tr} = \int_0^{+\infty} s_T(q) \frac{Q_0}{4\pi} \frac{e^{jk_2 R_T(q)}}{R_T(q)} dq, \quad (10.118)$$

where $R_T(q) = \sqrt{(x-x_q)^2 + (y-y_q)^2 + (z-H_T(q))^2}$. To ensure that the integral converges and ϕ_{tr} satisfies the radiation condition, we require that real and imaginary part of R_T be positive

$$\mathrm{Re}(R_T) > 0 \,\mathrm{and}\, \mathrm{Im}(R_T) > 0. \tag{10.119}$$

Equation (10.119) indicates that ϕ_{tr} can be represented by a distribution of point sources with density $s_T(q)$ located at positions $z_q(q) = H_T(q)$.

To find the source distribution $s_T(q)$, we analyze $\hat{T}_w(k_{2z})$. From (10.99), one can express k_{1z} in terms of k_{2z} as $k_{1z} = \sqrt{k_{2z}^2 - B^2}$. It can be shown that $\hat{T}_w(k_{2z})$ can be written as

$$\hat{T}_w(k_{2z}) = 2k_{2z}\left(\frac{1}{\alpha+1} - \frac{2}{\alpha^2-1}\sum_{n=1}^{\infty}\left(\frac{\alpha-1}{\alpha+1}\right)^n\left(\frac{B}{\sqrt{k_{2z}^2 - B^2} + k_{2z}}\right)^{2n}\frac{\mathrm{e}^{j\sqrt{k_{2z}^2-B^2}h}}{\sqrt{k_{2z}^2 - B^2}}\right). \tag{10.120}$$

The deduction is shown in Appendix D. To continue with the analysis, Lindell proposes to use the integral

$$\frac{\mathrm{e}^{-\sqrt{a^2+b^2}}}{\sqrt{a^2+b^2}}\left(\frac{b}{a+\sqrt{a^2+b^2}}\right)^{\nu} = \int_1^{+\infty}\left(\frac{x-1}{x+1}\right)^{\nu/2}\mathrm{e}^{-ax}J_\nu\left(b\sqrt{x^2-1}\right)\mathrm{d}x, \tag{10.121}$$

which is a Laplace transform with a and b real and positive. He assumes that it can be extended to complex variables. By setting

$$a = -jk_{2z}h, b = Bh \to \sqrt{a^2+b^2} = -jk_{1z}h, \frac{b}{a+\sqrt{a^2+b^2}} = \frac{jB}{k_{2z} + \sqrt{k_{2z}^2 - B^2}} \text{ and}$$

$$\frac{\mathrm{e}^{j\sqrt{k_{2z}^2-B^2}h}}{-j\sqrt{k_{2z}^2 - B^2}h}\left(\frac{jB}{k_{2z}+\sqrt{k_{2z}^2 - B^2}}\right)^{2n} = \int_0^{+\infty}\left(\frac{x-1}{x+1}\right)^n \mathrm{e}^{jk_{2z}x}J_{2n}\left(Bh\sqrt{x^2-1}\right)\mathrm{d}x. \tag{10.122}$$

Making the change of variable $q = h\sqrt{x^2-1}$, one has $x = H_T(q)/h$ with $H_T(q) = \sqrt{q^2+h^2}$ and (10.122) becomes

$$\frac{e^{j\sqrt{k_{2z}^2-B^2}h}}{\sqrt{k_{2z}^2-B^2}}\left(\frac{B}{k_{2z}+\sqrt{k_{2z}^2-B^2}}\right)^{2n} = -j\int_0^\infty\left(\frac{h-H_T(q)}{h+H_T(q)}\right)^n \cdot e^{jk_{2z}H_T(q)}J_{2n}(Bq)\frac{dH_T(q)}{dq}dq \quad (10.123)$$

For $\mathrm{Im}(B) = 0$, the integral on the right converges for $\mathrm{Im}(k_{2z}) > 0$. For a complex argument, the asymptotic value of $J_\nu(z)$ for $|z| \to \infty$ is [Abr70]

$$J_\nu(z) = \sqrt{\frac{2}{\pi z}}\cos\left(z - \frac{\nu\pi}{2} - \frac{\pi}{4}\right) + e^{|\mathrm{Im}(z)|}O\left(\frac{1}{|z|}\right), \quad |\arg(z)| < \pi,$$

i.e., $J_\nu(z)$ grows with $e^{|\mathrm{Im}(z)|}$. For $q \to \infty$, $H_T(q) \to q$, therefore, the integral converges only for the condition:

$$\mathrm{Im}(k_{2z}) > |\mathrm{Im}(B)| > 0. \quad (10.124)$$

Inserting (10.123) into (10.120), we have

$$\hat{T}_w(k_{2z}) = -jk_{2z}\int_0^\infty F(\alpha, B, q)e^{jk_{2z}H_T(q)}\frac{dH_T(q)}{dq}dq, \quad (10.125)$$

$$F(\alpha, B, q) = \frac{2}{\alpha+1}J_0(Bq) - \frac{4}{\alpha^2-1}\sum_{n=1}^\infty\left(\frac{\alpha-1}{\alpha+1}\cdot\frac{h-H_T(q)}{h+H_T(q)}\right)^n J_{2n}(Bq). \quad (10.126)$$

$F(\alpha, B, q)$ will converge only for $\mathrm{Im}(B) = 0$.

Applying the integration by parts in (10.125), we see that

$$\int_0^\infty F(\alpha, B, q)e^{jk_{2z}H(q)}\frac{dH_T(q)}{dq}dq = \left.\frac{F(\alpha, B, q)}{jk_{2z}}\right|_0^\infty - \frac{1}{jk_{2z}}\int_0^\infty e^{jk_{2z}H_T(q)}\frac{dF(\alpha, B, q)}{dq}dq. \quad (10.127)$$

For $q = 0$, $F(\alpha, B, q) = \frac{2}{\alpha+1}$ and $q \to \infty$, $F(\alpha, B, q) \to 0$ hence

$$\int_0^\infty F(\alpha, B, q)e^{jk_{2z}H_T(q)}\frac{dH_T(q)}{dq}dq = -\frac{1}{jk_{2z}}\int_0^\infty\left(\frac{2}{\alpha+1}\delta_+(q)\right.$$

$$+\frac{\mathrm{d}F(\alpha, B, q)}{\mathrm{d}q}\Bigg)\mathrm{e}^{jk_{2z}H_T(q)}\mathrm{d}q \tag{10.128}$$

Finally,

$$\hat{T}(k_{2z}) = \int\limits_0^{\infty}\left(\frac{2}{\alpha+1}\delta_+(q) + \frac{\mathrm{d}F(\alpha, B, q)}{\mathrm{d}q}\right)\mathrm{e}^{jk_{2z}H_T(q)}\mathrm{d}q \tag{10.129}$$

and the source density is

$$s_T(q) = \frac{2}{\alpha+1}\delta_+(q) + \frac{\mathrm{d}F(\alpha, B, q)}{\mathrm{d}q}. \tag{10.130}$$

Inserting (10.130) into (10.118), the transmitted wave is given by

$$\phi_{\mathrm{tr}} = \frac{2}{\alpha+1}\frac{Q_0\mathrm{e}^{jk_2R_1}}{4\pi R_1} + \int\limits_0^{+\infty}\frac{\mathrm{d}F(\alpha, B, q)}{\mathrm{d}q}\frac{Q_0}{4\pi}\frac{\mathrm{e}^{jk_2R_T(q)}}{R_T(q)}\mathrm{d}q. \tag{10.131}$$

Since the source locations are given by $z_q = H_T(q)$, i.e., they occupy the upper half-space, the source distribution is valid but only for a purely real B, namely for a real c_2 and for $c_1 > c_2$. The expression for $\frac{\mathrm{d}F(\alpha,B,q)}{\mathrm{d}q}$ is given explicitly in Appendix D.

Discussion on the formula (10.122)

In their paper from 1986, Lindell et al. propose the use of the integral (10.121) but they don't explicitly show how they perform the extension to the complex values. From their final formula, one can deduce that they make the change of variable

$$q = Bh\sqrt{x^2-1}$$

and when they determine the upper limit of integration, they write

$$\text{for}\, x \to \infty, q \to \infty$$

and the relation becomes

$$\frac{\mathrm{e}^{j\sqrt{k_{2z}^2-B^2}h}}{\sqrt{k_{2z}^2-B^2}}\left(\frac{B}{k_{2z}+\sqrt{k_{2z}^2-B^2}}\right)^{2n} = -j\int\limits_0^{\infty}\left(\frac{h-\tilde{H}_T(q)}{h+\tilde{H}_T(q)}\right)^n \mathrm{e}^{jk_{2z}\tilde{H}_T(q)}$$

$$J_{2n}(q)\frac{\mathrm{d}\tilde{H}_T(q)}{\mathrm{d}q}\mathrm{d}q \tag{10.132}$$

with $\tilde{H}_T(p) = \sqrt{(p/B)^2 + h^2}$. That is true only when $\varphi = \arg(B) = 0$ but when B is complex the upper limit should be

$$\text{for}\, x \to \infty, q \to \infty e^{j\varphi}, \tag{10.133}$$

i.e., the path of integration is not the real q-axis but a straight line in the complex plane that makes an angle φ with the real axis. Along this path, $J_{2n}(q) \to \infty$ for $|q| \to \infty$ unless $\varphi = 0$. So, Lindell et al. changed the integration path. Interestingly, in a book published in 1992 by Lindell [Lin92, Eq. 7.298], relation (10.122) is used instead of (10.132) but nothing is said about the divergence of the integral for $\varphi \neq 0$.

To see if the relation (10.132) is also correct, we performed a numerical study of the expression and compare the results of the integral with the exact values. By taking a closer look at the integrand in (10.132), we note that it converges for the values of k_{2z} that satisfy the relation

$$\mathrm{Re}(B)\mathrm{Im}(k_{2z}) > \mathrm{Im}(B)\mathrm{Re}(k_{2z}), \tag{10.134}$$

which is a less restrictive condition that (10.124) for the integral in (10.122). We do the same study for expression (10.122). We consider four different sets of values for k_{2z}. Figure 10.12 shows the comparison of the exact values (continuous lines) with the results of the integrals (circles).

The results from the integral (10.122) are plotted on the left and the results from the integral (10.132) on the right. In both cases, the results from the integral agree very good with the exact values when they can be computed. The integral (10.122) converges only for some values of each set while the integral (10.132) converges for more values of the four sets confirming that it is numerically less restricted. From the numerical study, we can infer the correctness of expression (10.132).

10.2.3 Comparison Between the Sommerfeld Approach and the EIT

(a) Reflected wave: In both approaches, the formulas are numerically stable for purely real or complex values of c_2. Figures 10.13 and 10.14 show the sound pressure at a set of field points in the upper half-space and the relative error of the EIT values by taking the Sommerfeld formula as the reference. The density and sound speed of medium 1 (air) are $\rho_1 = 1.21$ kg/m^3 and $c_1 = 343$ m/s. In Fig. 10.13, the density and sound speed of medium 2 are larger than in medium 1, $\rho_2 = 6.05$ kg/m^3 and $c_2 = 500$ m/s.

While the difference in the results from both approaches is very small, the difference in time of computation is considerable. The time of calculation with the Sommerfeld approach is just 0.1 s while the time of calculation with the EIT is 590 s. The

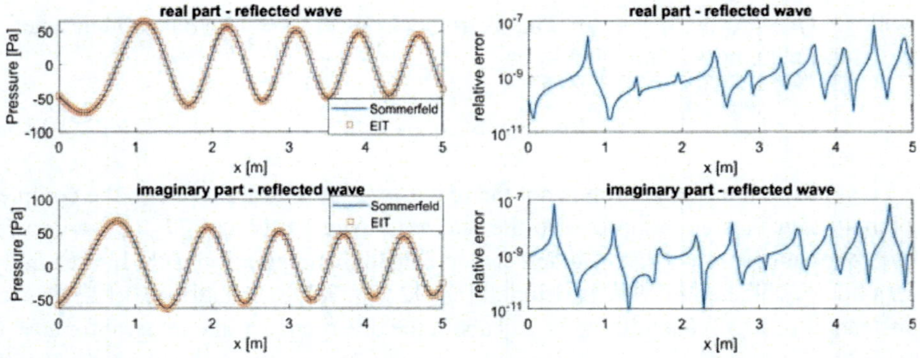

Fig. 10.13 Comparison of the sound pressure of the reflected wave obtained with the Sommerfeld approach and the EIT ($c_2 > c_1$)

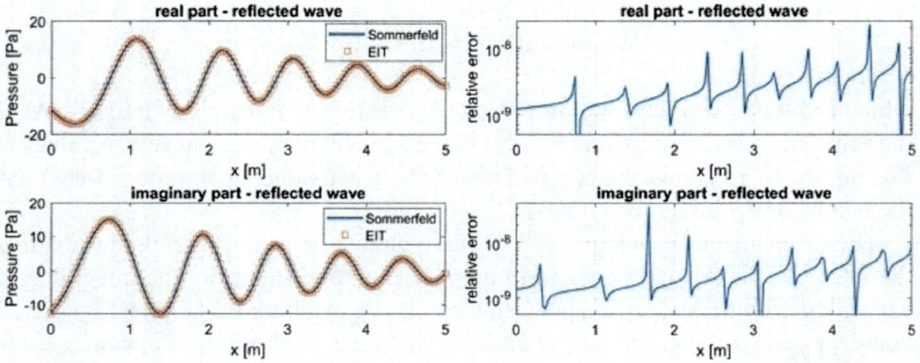

Fig. 10.14 Comparison of the sound pressure of the reflected wave obtained with the Sommerfeld approach and the EIT ($c_2 < c_1$)

EIT takes so long because $\mathrm{Re}(B) = 0$ and the equivalent source z_q is purely real, i.e., R_2^c is also purely real and the integrand in (10.118) does not have an exponential decaying term. A way to reduce the time of computation would be to include a small imaginary part in c_2, to provide the missing decaying term. For example, if we consider a complex sound speed $c_2(1 - 0.01j)$, the time of calculation will reduce to only 8 s, with an error below 1% in the sound pressure up to $x = 5$ m. However, this error will grow by increasing distance from the source.

In Fig. 10.14, the density and sound speed of medium 2 are $\rho_2 = 2.42$ kg/m^3 and $c_2 = 250$ m/s.

In the second case, the difference in the results but also the time of calculation from both approaches is very small. The time of calculation with the Sommerfeld approach is again about 0.1 s while the time of calculation with the EIT is 0.2 s. The EIT is very fast because $\mathrm{Im}(B) = 0$ and the equivalent source z_q is complex, i.e., R_2^c is also complex and the integrand in (10.118) has an exponential decaying term.

(b) Transmitted wave: The Sommerfeld formula is numerically stable for purely real or complex values of c_2. The EIT formula (10.131) is valid only for a purely real B, i.e., for a real c_2 and $c_2 < c_1$. Since (10.132) is also valid, we can insert it in the formula (10.120) and we obtain

$$\hat{T}_w(k_{2z}) = -jk_{2z} \int_0^\infty \tilde{F}(\alpha, B, q) \mathrm{e}^{jk_{2z}\tilde{H}_T(q)} \frac{\mathrm{d}\tilde{H}_T(q)}{\mathrm{d}q} \mathrm{d}q, \tag{10.135}$$

$$\tilde{F}(\alpha, B, q) = \frac{2}{\alpha+1} J_0(q) - \frac{4}{\alpha^2-1} \sum_{n=1}^{\infty} \left(\frac{\alpha-1}{\alpha+1} \cdot \frac{h - \tilde{H}_T(q)}{h + \tilde{H}_T(q)} \right)^n J_{2n}(q). \tag{10.136}$$

The source density is then

$$\tilde{s}_T(q) = \frac{2}{\alpha+1} \delta_+(q) + \frac{\mathrm{d}\tilde{F}(\alpha, B, q)}{\mathrm{d}q} \tag{10.137}$$

and the transmitted wave is given by

$$\phi_{\mathrm{tr}} = \frac{2}{\alpha+1} \frac{Q_0 e^{jk_2 R_1}}{4\pi R_1} + \int_0^{+\infty} \frac{\mathrm{d}\tilde{F}(\alpha, B, q)}{\mathrm{d}q} \frac{Q_0}{4\pi} \frac{\mathrm{e}^{jk_2 \tilde{R}_T(q)}}{\tilde{R}_T(q)} \mathrm{d}q. \tag{10.138}$$

In contrast to the integral in (10.131), the integral in (10.138) will contain a source density with complex source positions $z_q = \tilde{H}_T(q)$. In order to have a valid source density, it should not have a singularity in the lower half-space and should produce a complex distance with positive imaginary part. Both conditions are satisfied if

$$\mathrm{Re}(\tilde{H}_T(q)) \geq 0 \,\mathrm{and}\, \mathrm{Im}(\tilde{H}_T(q)) \geq 0. \tag{10.139}$$

Analyzing $\tilde{H}_T(q)$, we see that

$$\tilde{H}_T(u) = \sqrt{u^2 + h^2}, \quad u = \frac{q}{B}.$$

Since $\mathrm{Re}(B) > 0$ and $\mathrm{Im}(B) > 0 \rightarrow \mathrm{Re}(u) > 0$ and $\mathrm{Im}(u) < 0$ i.e., u belongs to the fourth quadrant. But in the fourth quadrant, real and imaginary parts of $\tilde{H}_T(q)$ have different signs, therefore (10.139) can be satisfied only when B is purely real or purely imaginary. That will be the case if c_2 is real. Hence, (10.138) has an advantage over (10.131), it converges also for the case $c_2 > c_1$. Unfortunately, it does not converge for complex c_2. When B is purely real, the source density is located in real positions with $z_q \in [h, \infty[$ and when B is purely imaginary, the source density is located partially in real positions and partially in complex positions with $z_q \in [0, h[\cup j[0, \infty[$.

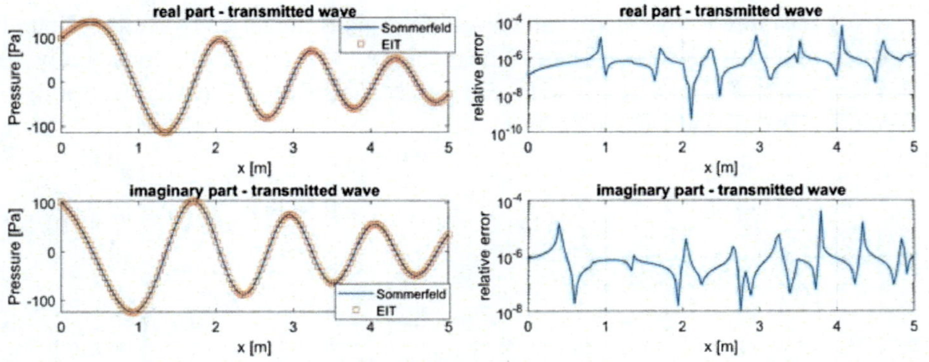

Fig. 10.15 Comparison of the sound pressure of the transmitted wave obtained with the Sommerfeld approach and the EIT ($c_2 > c_1$)

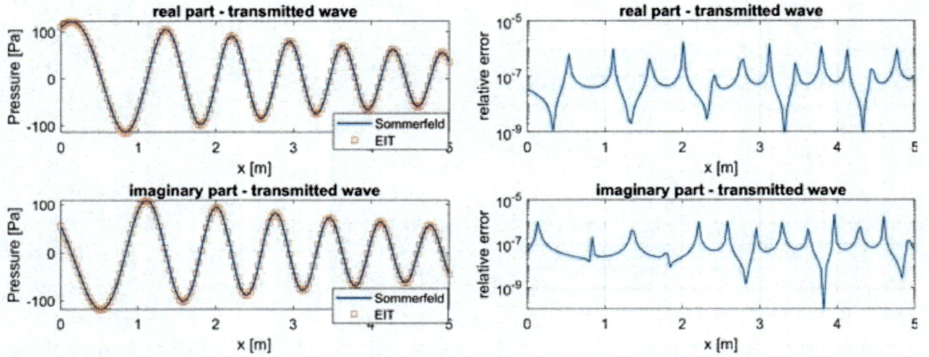

Fig. 10.16 Comparison of the sound pressure of the transmitted wave obtained with the Sommerfeld approach and the EIT ($c_2 < c_1$)

Figures 10.15 and 10.16 show the sound pressure at a set of field points in the lower half-space and the relative error of the EIT values by taking the Sommerfeld formula as the reference. Figure 10.15 presents the results for $c_2 > c_1$ and Fig. 10.16 for $c_2 < c_1$. In both cases, the difference in the sound pressure is negligible but the time of calculation is considerable larger with the EIT approach than with the Sommerfeld approach. For the case $c_2 > c_1$ ($\mathrm{Re}(B) = 0$), the time of calculation is 244 s and for the case $c_2 < c_1$ ($\mathrm{Im}(B) = 0$), 455 s, while the time of calculation with the Sommerfeld approach is just 0.2 s. For these calculations, the first 50 terms of the sum in $\frac{\mathrm{d}\hat{\hat{F}}(\alpha,B,q)}{\mathrm{d}q}$ were taken into account and the results indicate that it was enough.

Chapter 11
Boundary Element Techniques for Sound Propagation Above Impedance Planes

In this chapter, we study the implementation of Green's functions from previous chapters in the boundary element method (BEM) for simulating sound radiation and scattering above an infinite, flat plane with a locally reacting impedance. First, the basic concepts of the 3D half-space BEM are described, then some examples of application are provided.

11.1 Basic Concepts of the BEM for Half-Space Problems

The BEM is based on the discretization of an integral equation derived from the original partial differential equation. The integral equation is defined on the boundary of the domain and relates the solution on the boundary to the solution at points in the domain. In acoustics, the differential equation is the wave equation in time domain or the Helmholtz equation in frequency domain.

11.1.1 Direct Formulation

For the half-space problem, we distinguish the cases of radiation and scattering since the derivation of the integral equations differs [Sey88], [Och08a].

11.1.1.1 Sound Radiation

We consider a three-dimensional body with closed boundary Σ and outward normal vector n above an infinite plane Σ_p. The plane has a locally reactive impedance Z_p. The exterior of Σ is denoted by Ω (see Fig. 11.1).

M. Ochmann and R. Piscoya, *Theory and Application of Acoustic Sources Using Complex Analysis*, https://doi.org/10.1007/978-981-33-6040-2_11

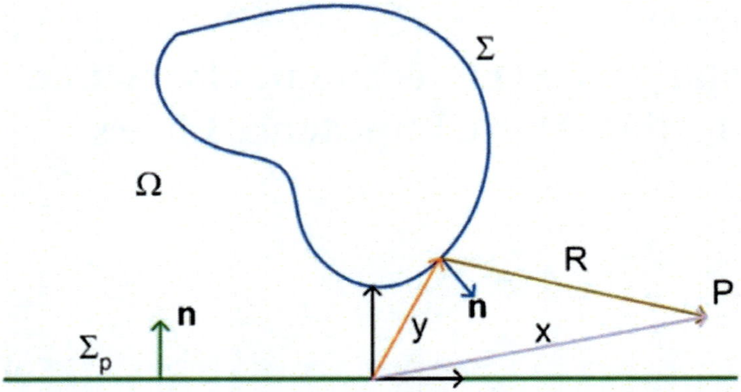

Fig. 11.1 Acoustic domain Ω and boundary surfaces Σ and Σ_p

For the radiation case, the sound pressure at every point x is obtained by expressing the Helmholtz equation in integral form

$$c(x)p(x) = \int\limits_{\Sigma\cup\Sigma_p} \left(p(y)\frac{\partial g(x,y)}{\partial n(y)} - \frac{\partial p(y)}{\partial n(y)} g(x,y) \right) \mathrm{d}S(y). \tag{11.1}$$

Here, $g(x, y)$ is any Green's function that satisfies the inhomogeneous Helmholtz equation:

$$\Delta g(x,y) + k^2 g(x,y) = -\delta(x-y),$$

where k is the wave number $k = \omega/c$ and the normal vector on the surface points toward the acoustic domain. For points in the acoustic domain, $c(x)$ has the value 1 and outside it has the value 0. For points on the surface, the constant $c(x)$ has the value 0.5 if the boundaries are smooth, i.e. if there exists only one unique tangent plane at each point. If not, $c(x)$ depends on the solid angle, which is formed by the surface Σ at point x. Obviously, (11.1) is not a very useful expression because the infinite plane is included in the integration domain and should be also discretized. Equation (11.1) shows explicitly the dependence on the variables x and y. From this point on, a shorter form without the dependence on the variables will be used.

Since the sound pressure satisfies the impedance boundary condition on the plane:

$$jkp - \frac{Z_p}{\rho c}\frac{\partial p}{\partial n} = 0.$$

Equation (11.1) can be rewritten as

$$cp = \int\limits_{\Sigma} \left(p\frac{\partial g}{\partial n_y} - \frac{\partial p}{\partial n_y} g \right) \mathrm{d}S + \int\limits_{\Sigma_p} \frac{j}{k}\frac{\partial p}{\partial n_y} \left(jkg - \frac{Z_p}{\rho c}\frac{\partial g}{\partial n_y} \right) \mathrm{d}S. \tag{11.2}$$

If we select a Green's function g^h that also satisfies the boundary condition on the plane:

$$jkg^h - \frac{Z_p}{\rho c}\frac{\partial g^h}{\partial n_y} = 0, \tag{11.3}$$

the second integral in (11.2) vanishes and the integration domain contains only the vibrating body.

$$cp = \int_{\Sigma} \left(p\frac{\partial g^h}{\partial n_y} - \frac{\partial p}{\partial n_y} g^h \right) \mathrm{d}S. \tag{11.4}$$

11.1.1.2 Sound Scattering

For the scattering case, the derivation of the integral equation is slightly more complicated. It is assumed that the total sound pressure has three components: the incident pressure (p^{inc}), its reflection on the infinite plane (p^{ref}), and the scattered pressure due to the presence of the body (p^s), $p = p^{\mathrm{inc}} + p^{\mathrm{ref}} + p^s$. The scattered pressure satisfies the integral Eq. (11.1) and the Sommerfeld radiation condition.

$$cp^s = \int_{\Sigma \cup \Sigma_p} \left(p^s\frac{\partial g}{\partial n_y} - \frac{\partial p^s}{\partial n_y} g \right) \mathrm{d}S. \tag{11.5}$$

For the interior volume enclosed by Σ, the sum of incident and reflected pressure $p^b = p^{\mathrm{inc}} + p^{\mathrm{ref}}$ satisfies the integral equation for interior problems (see [Och08a])

$$(1-c)p^b = -\int_{\Sigma} \left(p^b\frac{\partial g}{\partial n_y} - \frac{\partial p^b}{\partial n_y} g \right) \mathrm{d}S. \tag{11.6}$$

Subtracting (11.6) from (11.5), we obtain

$$cp = \int_{\Sigma} \left(p\frac{\partial g}{\partial n_y} - \frac{\partial p}{\partial n_y} g \right) \mathrm{d}S + \int_{\Sigma_p} \left(p^s\frac{\partial g}{\partial n_y} - \frac{\partial p^s}{\partial n_y} g \right) \mathrm{d}S + p^b. \tag{11.7}$$

In the absence of the scatterer, $p^s = 0$ and $p = p^b$, hence both p^s and p^b simultaneously satisfy the boundary condition. Therefore, the second integral can be written as in (11.3), and it can be again set to zero if we select a Green's function that also satisfies the boundary condition. The expression for the total sound pressure is then written as:

$$cp = \int_{\Sigma} \left(p \frac{\partial g^h}{\partial n_y} - \frac{\partial p}{\partial n_y} g^h \right) \mathrm{d}S + p^b. \tag{11.8}$$

The difference between (11.7) and (11.3) is the background pressure p^b in the absence of the scattering body.

11.1.2 Indirect Formulation

When the radiating or the scattering body is very thin, the direct formulation is numerically highly costly and unpractical. In this case, a more efficient formulation is required. The start point is the direct formulation. We consider again a closed surface, where the actual body is only a part of the whole body while the rest of the body can be assumed as a permeable fictitious surface. We define two regions, Ω^+ outside Σ and Ω^- inside Σ and divide Σ into two parts: $\Sigma = \Sigma_B \cup \Sigma_f$. In the region Ω^+, the sound pressure and its derivative are denoted as p^+ and $\partial p^+/\partial n$ and in region Ω^-, p^- and $\partial p^-/\partial n$.

11.1.2.1 Sound Radiation

In region Ω^+, the integral equation is given by

$$cp^+ = \int_{\Sigma_B} \left(p^+ \frac{\partial g^h}{\partial n_y} - \frac{\partial p^+}{\partial n_y} g^h \right) \mathrm{d}S + \int_{\Sigma_f} \left(p^+ \frac{\partial g^h}{\partial n_y} - \frac{\partial p^+}{\partial n_y} g^h \right) \mathrm{d}S. \tag{11.9}$$

In region Ω^-, any Green's function can be used, since it is not coupled to the ground, but using the half-space Green's function is useful, so the integral equation is written as

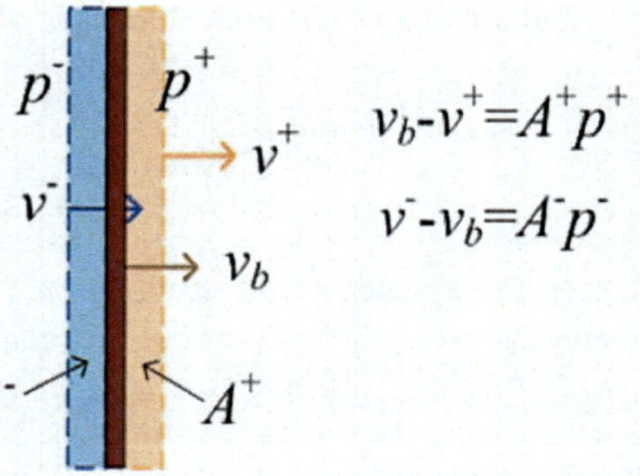

Fig. 11.2 A thin radiating surface lined with absorbing material

$$cp^- = -\int_{\Sigma_B} \left(p^- \frac{\partial g^h}{\partial n_y} - \frac{\partial p^-}{\partial n_y} g^h \right) \mathrm{d}S - \int_{\Sigma_f} \left(p^- \frac{\partial g^h}{\partial n_y} - \frac{\partial p^-}{\partial n_y} g^h \right) \mathrm{d}S. \tag{11.10}$$

By adding (11.10) and (11.11), we obtain the equation

$$c\tilde{p} = \int_{\Sigma_B} \left(\Delta p \frac{\partial g^h}{\partial n_y} - \Delta\sigma g^h \right) \mathrm{d}S + \int_{\Sigma_f} \left(\Delta p \frac{\partial g^h}{\partial n_y} - \Delta\sigma g^h \right) \mathrm{d}S,$$

where $\tilde{p} = p^+ + p^-$, $\Delta p = p^+ - p^-$, and $\Delta\sigma = \frac{\partial p^+}{\partial n} - \frac{\partial p^-}{\partial n}$. Finally, since Σ_f is a fictitious permeable surface, $\Delta p = 0$, $\Delta\sigma = 0$ in the second integral and the indirect formulation simplifies to

$$c\tilde{p} = \int_{\Sigma_B} \left(\Delta p \frac{\partial g^h}{\partial n_y} - \Delta\sigma g^h \right) \mathrm{d}S. \tag{11.11}$$

In this formulation, the important variables on the surface are the differences of pressure and their derivatives Δp and $\Delta\sigma$, respectively, instead of the quantities itself. The direction of the normal vector is for this approach irrelevant. It only serves to denote a positive side and a negative side of the surface.

On the surface, integral equation (11.11) is not very useful since we have one equation and two unknowns $\bar{p}$ and Δp. To solve the problem, the normal derivative of (11.11) is performed and we obtain a second equation

$$c\tilde{\sigma} = \int_{\Sigma_B} \left(\Delta p \frac{\partial^2 g^h}{\partial n_x \partial n_y} - \Delta\sigma \frac{\partial g^h}{\partial n_x} \right) \mathrm{d}S, \tag{11.12}$$

where $\tilde{\sigma} = \frac{\partial p^+}{\partial n_x} + \frac{\partial p^-}{\partial n_x}$.

By radiation problems, usually the normal velocity of the surface v_b is known and it is equal an either side of the surface, i.e., $\Delta\sigma = 0$. On the surface, $c\tilde{\sigma} = -j\rho\omega v_b$ and (11.12) is simplified to

$$-j\rho\omega v_b = \int_{\Sigma_B} \Delta p \frac{\partial^2 g^h}{\partial n_x \partial n_y} \mathrm{d}S. \tag{11.13}$$

In some applications, the vibrating surface can be lined by an absorbing material (see Fig. 11.2).

Assuming that the liners are different at each side of the surface and characterized by the admittances A^+ and A^-, the velocities of the fluid at the exterior surface of the absorbers are given by

$$\begin{array}{l} v_b - v^+ = A^+ p^+ \\ v^- - v_b = A^- p^- \end{array} \quad \rightarrow \quad \begin{array}{l} \Delta v = v^+ - v^- = -A^+ p^+ - A^- p^- \\ \tilde{v} = v^+ + v^- = 2v_b - A^+ p^+ + A^- p^- \end{array}. \tag{11.14}$$

By introducing the definitions $\tilde{A}_0 = \rho c(A^+ + A^-)$ and $\Delta A_0 = \rho c(A^+ - A^-)$, we can define $\tilde{\sigma}$ and $\Delta\sigma$ in terms of $\tilde{p}$ and Δp as follows

$$\begin{aligned} \tilde{\sigma} &= \frac{\tilde{A}_0}{2} jk\Delta p + \frac{\Delta A_0}{2} jk\tilde{p} - j2\rho\omega v_b \\ \Delta\sigma &= \frac{\tilde{A}_0}{2} jk\tilde{p} + \frac{\Delta A_0}{2} jk\Delta p. \end{aligned} \tag{11.15}$$

Inserting (11.15) into (11.11) and (11.12), we obtain a set of two differential equations on the surface with two unknowns

$$\begin{aligned} &\frac{1}{2} jk\tilde{p} + \int_{\Sigma_B} jk\tilde{p}\frac{\tilde{A}_0}{2} jkg^h \mathrm{d}S - \int_{\Sigma_B} jk\Delta p\left(\frac{\partial g^h}{\partial n_y} - \frac{\Delta A_0}{2} jkg^h\right)\mathrm{d}S = 0 \\ &\frac{\Delta A_0}{4} jk\tilde{p} + \int_{\Sigma_B} jk\tilde{p}\frac{\tilde{A}_0}{2}\frac{\partial g^h}{\partial n_y}\mathrm{d}S + \frac{\tilde{A}_0}{4} jk\Delta p \\ &\quad + \int_{\Sigma_B} jk\Delta p\left(\frac{j}{k}\frac{\partial^2 g^h}{\partial n_x \partial n_y} + \frac{\Delta A_0}{2}\frac{\partial g^h}{\partial n_x}\right)\mathrm{d}S = j\rho\omega v_b. \end{aligned} \tag{11.16}$$

11.1.2.2 Sound Scattering

The integral equations at both sides of Σ are

$$cp^+ = \int_{\Sigma_B} \left(p^+\frac{\partial g^h}{\partial n_y} - \frac{\partial p^+}{\partial n_y} g^h\right)\mathrm{d}S + \int_{\Sigma_f} \left(p^+\frac{\partial g^h}{\partial n_y} - \frac{\partial p^+}{\partial n_y} g^h\right)\mathrm{d}S + p_b, \tag{11.17}$$

$$cp^- = -\int_{\Sigma_B} \left(p^-\frac{\partial g^h}{\partial n_y} - \frac{\partial p^-}{\partial n_y} g^h\right)\mathrm{d}S - \int_{\Sigma_f} \left(p^-\frac{\partial g^h}{\partial n_y} - \frac{\partial p^-}{\partial n_y} g^h\right)\mathrm{d}S. \tag{11.18}$$

The incident wave in absence of the scatterer appears only in region Ω^+. By adding (11.17) and (11.18), we obtain the equation

$$c\tilde{p} = \int_{\Sigma_B} \left(\Delta p\frac{\partial g^h}{\partial n_y} - \Delta\sigma g^h\right)\mathrm{d}S + p^b. \tag{11.19}$$

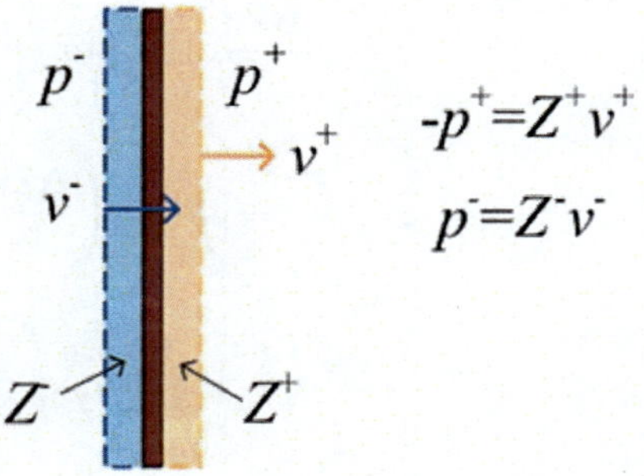

Fig. 11.3 A thin scattering surface covered with absorbing material

Performing the normal derivative of (11.19), we obtain

$$c\tilde{\sigma} = \int_{\Sigma_B} \left(\Delta p \frac{\partial^2 g^h}{\partial n_x \partial n_y} - \Delta\sigma \frac{\partial g^h}{\partial n_x} \right) \mathrm{d}S + \frac{\partial p^b}{\partial n_x}. \tag{11.20}$$

In scattering problems, usually the scatterer is rigid, i.e., $\tilde{\sigma} = 0$ and $\Delta\sigma = 0$, so that (11.20) is enough to solve the problem. It simplifies to

$$\int_{\Sigma_B} \Delta p \frac{\partial^2 g^h}{\partial n_x \partial n_y} \mathrm{d}S = -\frac{\partial p^b}{\partial n_x}. \tag{11.21}$$

There are however cases where the surface of the scatterer is covered with some absorbing material (see Fig. 11.3).

Assuming that the absorbing materials are different at each side of the surface and characterized by the impedances Z^+ and Z^-, at the exterior surface of the absorbers we have

$$\begin{array}{l} -p^+ = Z^+ v^+ \\ p^- = Z^- v^- \end{array} \quad \rightarrow \quad \begin{array}{l} \Delta p = p^+ - p^- = -Z^+ v^+ - Z^- v^- \\ \tilde{p} = p^+ + p^- = -Z^+ v^+ + Z^- v^- \end{array}. \tag{11.22}$$

By introducing the definitions $\tilde{Z}_0 = (Z^+ + Z^-)/\rho c$ and $\Delta Z_0 = (Z^+ - Z^-)/\rho c$, we can define $\tilde{p}$ and Δp in terms of $\tilde{\sigma}$ and $\Delta\sigma$ as follows

$$\begin{aligned} jk\tilde{p} &= \frac{\tilde{Z}_0}{2}\Delta\sigma + \frac{\Delta Z_0}{2}\tilde{\sigma} \\ jk\Delta p &= \frac{\tilde{Z}_0}{2}\tilde{\sigma} + \frac{\Delta Z_0}{2}\Delta\sigma. \end{aligned} \tag{11.23}$$

Inserting (11.23) into (11.19) and (11.20), we obtain a set of two differential equations on the surface with two unknowns

$$
\begin{aligned}
&\frac{\Delta Z_0}{4}\tilde{\sigma} - \int\limits_{\Sigma_B} \tilde{\sigma}\frac{\tilde{Z}_0}{2}\frac{\partial g^h}{\partial n_y}\mathrm{d}S + \frac{\tilde{Z}_0}{4}\Delta\sigma \\
&\quad - \int\limits_{\Sigma_B} \Delta\sigma\left(\frac{\Delta Z_0}{2}\frac{\partial g^h}{\partial n_y} - jkg^h\right)\mathrm{d}S = jkp^b \\
&\frac{1}{2}\tilde{\sigma} + \int\limits_{\Sigma_B} \tilde{\sigma}\frac{\tilde{Z}_0}{2}\frac{j}{k}\frac{\partial^2 g^h}{\partial n_x \partial n_y}\mathrm{d}S \\
&\quad + \int\limits_{\Sigma_B} \Delta\sigma\left(\frac{\Delta Z_0}{2}\frac{j}{k}\frac{\partial^2 g^h}{\partial n_x \partial n_y} + \frac{\partial g^h}{\partial n_x}\right)\mathrm{d}S = \frac{\partial p^b}{\partial n_x}.
\end{aligned}
\tag{11.24}
$$

11.2 Discretization

The numerical implementation of the BEM requires the discretization of the surface of the body and the corresponding integral equations. The surface is divided into a certain number of elements. The acoustical quantities are then computed at a finite set of points on the element and the value at any other point within the element is interpolated from those values using the so-called shape functions. Usually the shape functions are polynomials of certain degree. The simplest one is the polynomial of zero-th order. For that case, the acoustic quantities are computed at the center of the element and are constant in the whole surface of an element, which give rise to discontinuous elements. For higher orders of the polynomial, the quantities are computed at several points and they vary within the element which leads to continuous elements (see Fig. 11.4).

Using constant elements has the advantage that there is a uniquely defined normal vector in all elements (for higher orders the normal vectors are not well defined on the edges) and that for infinitely thin bodies, no extra condition on the edges and junctions (zero jump) is needed [Vla00]. For half-space BEM, there is the additional advantage, namely, for bodies sitting on the plane, source and receiving points do not lie on the ground. When that is the case, it can lead to some singularities on the

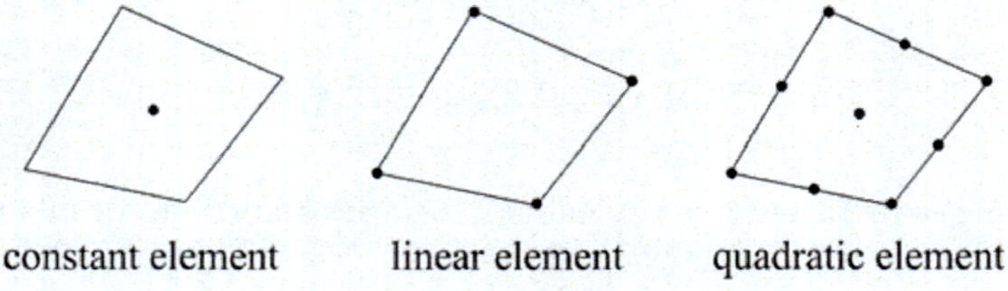

Fig. 11.4 Quadrilateral elements of different orders

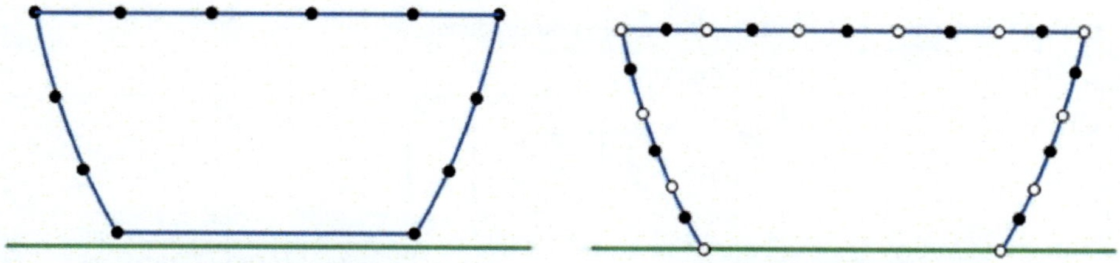

Fig. 11.5 Different discretization approaches for the half-space BEM

Green's function. This can be avoided by placing the body slightly above the ground (see Fig. 11.5).

If the body surface is divided into Q constant elements (E_j), the size of the resulting system of equations will be $Q \times Q$. Accurate results are expected if the discretization provides at least six points per wavelength.

After the discretization, Eq. (11.8) can be written in matrix form as

$$\left(\frac{1}{2}I - D^h\right)p + S^h\frac{\partial p}{\partial n} = p^b, \tag{11.25}$$

where I is the identity matrix and S^h and D^h are $Q \times Q$ matrices whose coefficients represent the interaction between two surface elements, defined as:

$$S^h_{ij} = \int_{E_j} g^h \mathrm{d}S, \quad D^h_{ij} = \int_{E_j} \frac{\partial g^h}{\partial n_y} \mathrm{d}S.$$

11.3 Non-uniqueness at Certain Frequencies

For the direct formulation, the solutions of the surface integral Eqs. (11.4) and (11.8) are not unique at a number of critical frequencies corresponding to internal resonances of the object. As the wave number k increases, so does the density of critical values, and hence it becomes increasingly difficult to get an accurate solution.

To overcome this problem, three approaches are available: the CHIEF method [Sch68], [Moh11a], the Burton and Miller method [Bur71], and the method of Dual Surface Integral Equations [Moh11b]. In the CHIEF method, the surface integral is evaluated in a certain number of interior points where the value is known to be zero. As a result, an over-determined system of equations has to be solved. In the Burton and Miller method, one uses a linear combination of the integral equation (11.4) or (11.8) and its first derivative. The coupling parameter γ has to have a nonzero imaginary part to ensure the uniqueness of the solution. Interestingly, Marburg [Mar16] has

Table 11.1 Comparison of methods to avoid non-uniqueness

Method	Matrix equation
CHIEF	$\begin{bmatrix} \frac{1}{2}I - D^h & S^h \\ -\bar{D}^h & \bar{S}^h \end{bmatrix} \begin{bmatrix} p \\ \frac{\partial p}{\partial n} \end{bmatrix} = \begin{bmatrix} p^b \\ \bar{p}^b \end{bmatrix}$
Burton and Miller	$(\frac{1}{2}I - D^h - \gamma M^h)p + (S^h + \gamma(\frac{1}{2}I + K^h))\frac{\partial p}{\partial n} = p^b + \gamma \frac{\partial p^b}{\partial n}$
Dual surface	$\left(\frac{1}{2}I - D^h - \alpha \bar{D}^h\right)p + \left(S^h + \alpha \bar{S}^h\right)\frac{\partial p}{\partial n} = p^b + \alpha \bar{p}^b$

shown that the "correct" sign of γ depends on the time convention chosen ($e^{-j\omega t}$ or $e^{+j\omega t}$). In the dual surface method, the integral equation evaluated on the surface is combined with the one evaluated at an interior surface with a purely imaginary coupling factor α to ensure the uniqueness. K^h and M^h are $Q \times Q$ matrices whose coefficients are defined as

$$K_{ij}^h = \int_{E_j} \frac{\partial g^h}{\partial n_x} \mathrm{d}S, \quad M_{ij}^h = \int_{E_j} \frac{\partial^2 g^h}{\partial n_x \partial n_y} \mathrm{d}S.$$

A comparison of the three methods is shown in Table 11.1. The matrices with an upper bar denote the matrices evaluated at the interior points. For more details, see [Moh11b].

11.4 Green's Function of Rigid ($Z_p \to \infty$) and Soft ($Z_p = 0$) Infinite Plane

For a time convention $e^{j\omega t}$, the free-space Green's function is given by:

$$g_F(x, y) = \frac{e^{-jkR}}{4\pi R}, \tag{11.26}$$

with $R = |x - y|$, $j^2 = -1$.

In the half-space, for a rigid or soft infinite plane, the Green's function can be constructed either by considering (a) an image source point (y^s) or (b) an image field point (P^s) (see Fig. 11.6). The second representation is useful to perform the derivative of g^h with respect to y. Introducing a reflection factor R_p, the Green' function is defined as:

$$g^h = \frac{e^{-jkR}}{4\pi R} + R_p \frac{e^{-jkR^s}}{4\pi R^s}, \tag{11.27}$$

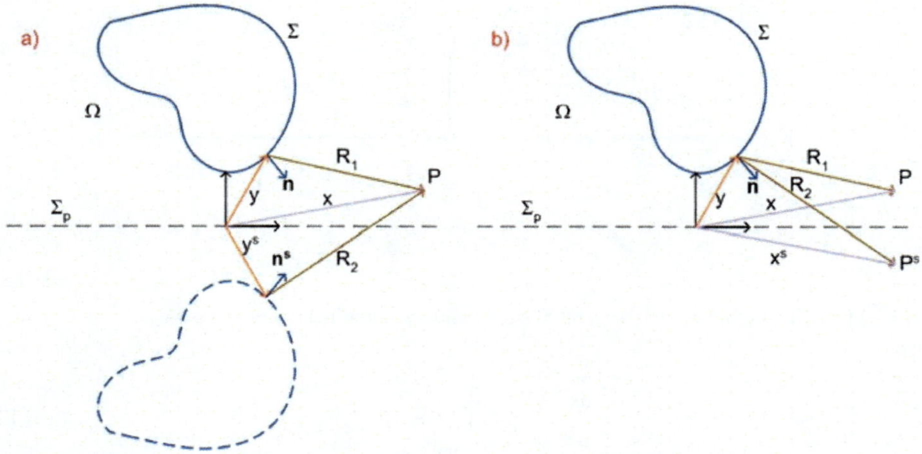

Fig. 11.6 Image source (y^s) or image field point (P^s) for the Green's function

where $R_p = 1$ if the plane is rigid or $R_p = -1$ if the plane is soft, $R = |x - y|$ and $R^s = |x - y^s| = |x^s - y|$.

For the scattering case, if the incident pressure is due to a plane wave instead of a point source,

$$p^b = p_0(\mathrm{e}^{-jkn_w \cdot x} + R_p \mathrm{e}^{-jkn_w^s \cdot x}),$$

where n_w and n_w^s are the direction of incident and reflected plane waves, respectively.

11.5 The General Impedance Boundary Condition at the Infinite Plane

Generally, the Green's function above an infinite plane with a surface impedance Z_p (g^h) can be written as the Green's function for the rigid plane (g_R^h) plus a correction term $C(\gamma)$

$$g^h = g_R^h + C(\gamma), \tag{11.28}$$

$$g_R^h = g^f + g^s, \quad g^f = \frac{\mathrm{e}^{-jkR}}{4\pi R}, \quad g^s = \frac{\mathrm{e}^{-jkR^s}}{4\pi R^s}, \tag{11.29}$$

where $\gamma = jk/Z_0$ and Z_0 is the normalized impedance $Z_0 = Z_p/\rho c$. Considering the infinite plane at the XY plane, with a source point placed at $\vec{y} = (y_1, y_2, h)$, the image source point is located at $\vec{y}^s = (y_1, y_2, -h)$ (see Fig. 11.7).

The correction term $C(\gamma)$ satisfies the condition

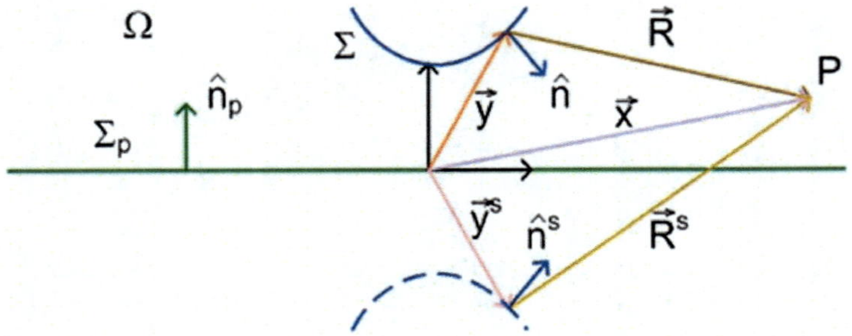

Fig. 11.7 Illustration of source point $\vec{y}$, field point $\vec{x}$, and mirror source point $\vec{y}^s$

$$\left[\frac{\partial C}{\partial n} - \gamma C\right]_{z=0} = \gamma\, g_R^h\big|_{z=0}, \tag{11.30}$$

which is obtained by inserting (11.28) into (11.3).

An efficient BEM implementation requires a representation of the Green's function that is numerically stable and also valid for all type of ground impedances. In the literature, one can find several formulations but not all of them possess the conditions previously mentioned. In Chap. 6 and in [Pis17] and [Pis19], several formulas have been studied. Here, we will present four of them that can be included in a BEM implementation. Explicit expressions of the first and second derivatives of the studied formulas are included, since they are necessary for the BEM formulation.

11.5.1 Classical Formula with Cylindrical Waves (Sommerfeld)

The formula given by Sommerfeld [Som78, p. 228, Eq. (9)] is the starting point of several representations including the formula of Thomasson (Sect. 11.5.2). In that representation, the correction term is given by

$$C(\gamma) = -\frac{\gamma}{2\pi}\int_0^\infty J_0(\lambda R_0)\mathrm{e}^{-\mu(x_3+h)}\frac{\lambda \mathrm{d}\lambda}{\mu(\mu+\gamma)}, \quad \mu = \sqrt{\lambda^2 - k^2}, \tag{11.31}$$

where $\vec{R}_0 = (x_1 - y_1, x_2 - y_2, 0)$, $R_0 = \left|\vec{R}_0\right|$.

By taking the proper value of the square root $\mu = \begin{cases} j\sqrt{k^2 - \lambda^2}, & \lambda < k \\ \sqrt{\lambda^2 - k^2}, & \lambda > k \end{cases}$, $C(\gamma)$ is written as

$$C(\gamma) = \frac{\gamma}{2\pi}\left(I_S^{(1)} - I_S^{(2)}\right), \tag{11.32}$$

$$I_S^{(1)} = \int_0^k J_0(\lambda R_0) \mathrm{e}^{-j\sqrt{k^2-\lambda^2}(x_3+h)} \frac{\lambda \mathrm{d}\lambda}{\sqrt{k^2-\lambda^2}\left(\sqrt{k^2-\lambda^2} - j\gamma\right)}, \tag{11.33}$$

$$I_S^{(2)} = \int_k^\infty J_0(\lambda R_0) e^{-\sqrt{\lambda^2-k^2}(x_3+h)} \frac{\lambda \mathrm{d}\lambda}{\sqrt{\lambda^2-k^2}\left(\sqrt{\lambda^2-k^2} + \gamma\right)}. \tag{11.34}$$

By performing the change of variables: $s = \sqrt{1-(\lambda/k)^2}$ in (11.42) and $s = \sqrt{(\lambda/k)^2 - 1}$ in (11.43), we obtain

$$I_S^{(1)} = \int_0^1 J_0(\sqrt{1-s^2}kR_0) \mathrm{e}^{-jks(x_3+h)} \frac{\mathrm{d}s}{s+\beta}, \tag{11.35a}$$

$$I_S^{(2)} = \int_0^\infty J_0(\sqrt{1+s^2}kR_0) \mathrm{e}^{-ks(x_3+h)} \frac{\mathrm{d}s}{s+j\beta}, \tag{11.36a}$$

where $\beta = \rho c / Z_p$.

The derivatives of (11.35a) and (11.36a) are given by

$$\begin{aligned} \frac{\partial I_S^{(1)}}{\partial n_x} = -k \int_0^1 & \Big[js(\hat{n}_p \cdot \hat{n}_x) J_0\left(\sqrt{1-s^2}kR_0\right) \\ & + \sqrt{1-s^2} J_1\left(\sqrt{1-s^2}kR_0\right) \frac{\partial R_0}{\partial n_x} \Big] \mathrm{e}^{-jks(x_3+h)} \frac{\mathrm{d}s}{s+\beta}, \end{aligned} \tag{11.35b}$$

$$\begin{aligned} \frac{\partial I_S^{(1)}}{\partial n_y} = k \int_0^1 & \Big[js(\hat{n}_p \cdot \hat{n}_y^s) J_0\left(\sqrt{1-s^2}kR_0\right) \\ & - \sqrt{1-s^2} J_1\left(\sqrt{1-s^2}kR_0\right) \frac{\partial R_0}{\partial n_y} \Big] \mathrm{e}^{-jks(x_3+h)} \frac{\mathrm{d}s}{s+\beta}, \end{aligned} \tag{11.35c}$$

$$\begin{aligned} \frac{\partial^2 I_S^{(1)}}{\partial n_x \partial n_y} = k \int_0^1 & \Big[kJ_0\left(\sqrt{1-s^2}kR_0\right) \left\{ s^2 (\hat{n}_p \cdot \hat{n}_x)(\hat{n}_p \cdot \hat{n}_y^s) - (1-s^2) \frac{\partial R_0}{\partial n_x} \frac{\partial R_0}{\partial n_y} \right\} \\ & - \sqrt{1-s^2} J_1\left(\sqrt{1-s^2}kR_0\right) \Big\{ \frac{\partial^2 R_0}{\partial n_x \partial n_y} - \frac{1}{R_0} \frac{\partial R_0}{\partial n_x} \frac{\partial R_0}{\partial n_y} \\ & + jks \left[(\hat{n}_p \cdot \hat{n}_y^s) \frac{\partial R_0}{\partial n_x} - (\hat{n}_p \cdot \hat{n}_x) \frac{\partial R_0}{\partial n_y} \right] \Big\} \Big] \mathrm{e}^{-jks(x_3+h)} \frac{\mathrm{d}s}{s+\beta}, \end{aligned} \tag{11.35d}$$

$$\frac{\partial I_S^{(2)}}{\partial n_x} = -k\int_0^\infty \Big[s(\hat{n}_p\cdot\hat{n}_x)J_0\big(\sqrt{1+s^2}kR_0\big) + \sqrt{1+s^2}J_1\big(\sqrt{1+s^2}kR_0\big)\frac{\partial R_0}{\partial n_x}\Big]\mathrm{e}^{-ks(x_3+h)}\frac{\mathrm{d}s}{s+j\beta}. \tag{11.36b}$$

$$\frac{\partial I_S^{(2)}}{\partial n_y} = k\int_0^\infty \Big[s(\hat{n}_p\cdot\hat{n}_y^s)J_0\big(\sqrt{1+s^2}kR_0\big) - \sqrt{1+s^2}J_1\big(\sqrt{1+s^2}kR_0\big)\frac{\partial R_0}{\partial n_y}\Big]\mathrm{e}^{-ks(x_3+h)}\frac{\mathrm{d}s}{s+j\beta}. \tag{11.36c}$$

$$\begin{aligned}\frac{\partial^2 I_S^{(2)}}{\partial n_x\partial n_y} = -k\int_0^\infty \Big[&kJ_0\big(\sqrt{1+s^2}kR_0\big)\Big\{s^2(\hat{n}_p\cdot\hat{n}_x)(\hat{n}_p\cdot\hat{n}_y^s) \\ &+(1+s^2)\frac{\partial R_0}{\partial n_x}\frac{\partial R_0}{\partial n_y}\Big\} \\ &+\sqrt{1+s^2}J_1\big(\sqrt{1+s^2}kR_0\big)\Big\{\frac{\partial^2 R_0}{\partial n_x\partial n_y} - \frac{1}{R_0}\frac{\partial R_0}{\partial n_x}\frac{\partial R_0}{\partial n_y} \\ &+ks\Big[(\hat{n}_p\cdot\hat{n}_y^s)\frac{\partial R_0}{\partial n_x} - (\hat{n}_p\cdot\hat{n}_x)\frac{\partial R_0}{\partial n_y}\Big]\Big\}\Big]\mathrm{e}^{-ks(x_3+h)}\frac{\mathrm{d}s}{s+j\beta},\end{aligned} \tag{11.36d}$$

The derivatives of R_0 are given in Appendix E.

Numerical stability

The integrals (11.35) have no singularities since $\mathrm{Re}(\beta) \geq 0$ and $|\beta| > 0$. The integrals (11.36) converge for $x_3 + h > 0$ and have a singularity for impedances with $\mathrm{Re}(\beta) = 0$ and $\mathrm{Im}(\beta) > 0$, i.e., for pure imaginary spring-like impedances. The singularity is integrable as it was shown in [Pis19], and numerically, the integrals are computed by inserting a semi-circular contour in the upper half of the s-complex plane with radio ε centered on the singular point (see Fig. 11.8).

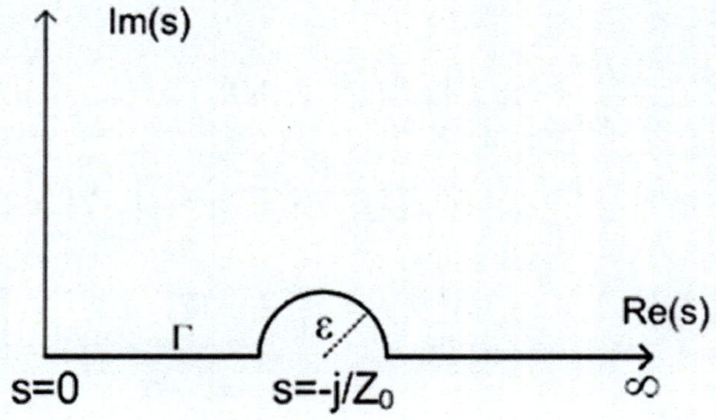

Fig. 11.8 Deformed integration path for integrals (11.36)

For $x_3 + h \ll 1$ and $R_0 \gg 1$, the integrand oscillates strongly due to the Bessel function and it will be difficult to approximate numerically.

11.5.2 Formula with Reflected Plane Waves (Thomasson)

Ingard [Ing51] used the decomposition of a point source in an infinitely number of plane waves given by Weyl [Wey19] to define the reflected sound pressure. Thomasson [Tho76] showed that in Ingard's solution, the "surface wave" term was missing. That term appears precisely by the spring-like impedances. The formula of Thomasson reads:

$$C(\gamma) = \begin{cases} \psi_{SD} + \psi_B, & \mathrm{Re}(\sigma) > 1, \quad \mathrm{Im}(\beta) > 0 \\ \psi_{SD}, & \text{else} \end{cases}, \tag{11.37}$$

where $\beta = \rho c/Z$, $\delta = (x_3 + h)/R^s$ and $\sigma = -\beta\delta + \sqrt{(1-\beta^2)(1-\delta^2)}$.

$$\psi_{\mathrm{SD}} = \frac{j\gamma}{2\pi} \mathrm{e}^{-jkR^s} I_W \tag{11.38a}$$

where

$$I_W = \int_0^\infty \frac{\mathrm{e}^{-kR^s t}}{\sqrt{W}} \mathrm{d}t, \quad \mathrm{Re}(\sqrt{W})$$

$$= \begin{cases} < 0, \ \mathrm{Re}(\sigma) > 1, & \mathrm{Im}(\beta) > 0, \quad t > t_1 \\ > 0, \ \text{else} \end{cases},$$

$$W = (\beta + \delta)^2 - j2t(1 + \beta\delta) - t^2, \quad t_1 = \frac{\mathrm{Im}(\beta)(\mathrm{Re}(\beta) + \delta)}{1 + \mathrm{Re}(\beta)\delta}.$$

To evaluate I_W, the proper branch cut of $\sqrt{W}$ has to be selected. By doing that, the right values of $\mathrm{Re}(\sqrt{W})$ as specified above are obtained. A detailed analysis of the integral contour was given by Thomasson in his paper.

The derivatives of (11.38a) are given by

$$\frac{\partial \psi_{\mathrm{SD}}}{\partial n_x} = -jk\psi_{\mathrm{SD}} \frac{\partial R^s}{\partial n_x} + \frac{j\gamma}{2\pi} \mathrm{e}^{-jkR^s} \frac{\partial I_W}{\partial n_x}, \tag{11.38b}$$

$$\frac{\partial \psi_{\mathrm{SD}}}{\partial n_y} = -jk\psi_{\mathrm{SD}} \frac{\partial R^s}{\partial n_y} + \frac{j\gamma}{2\pi} \mathrm{e}^{-jkR^s} \frac{\partial I_W}{\partial n_y}, \tag{11.38c}$$

$$\frac{\partial^2 \psi_{\mathrm{SD}}}{\partial n_x \partial n_y} = -jk\left(\psi_{\mathrm{SD}} \frac{\partial^2 R^s}{\partial n_x \partial n_y} + \frac{\partial \psi_{\mathrm{SD}}}{\partial n_x} \frac{\partial R^s}{\partial n_y}\right)$$

$$+\frac{j\gamma}{2\pi}e^{-jkR^s}\left(\frac{\partial^2 I_W}{\partial n_x \partial n_y}-jk\frac{\partial I_W}{\partial n_y}\frac{\partial R^s}{\partial n_x}\right), \tag{11.38d}$$

$$\psi_B=\frac{j\gamma}{2}H_0^{(2)}\left(kR_0\sqrt{1-\beta^2}\right)e^{jk\beta(x_3+h)}, \tag{11.39a}$$

with

$$\vec{R}_0=(x_1-y_1,x_2-y_2,0),\quad R_0=\left|\vec{R}_0\right|,\quad \mathrm{Re}\left(\sqrt{1-\beta^2}\right)>0.$$

The derivatives of (11.39a) are given by

$$\frac{\partial\psi_B}{\partial n_x}=jk\beta(\hat{n}_p\cdot\hat{n}_x)\psi_B-\frac{j\gamma}{2\pi}k\pi\sqrt{1-\beta^2}$$
$$H_1^{(2)}\left(kR_0\sqrt{1-\beta^2}\right)e^{jk\beta(x_3+h)}\frac{\partial R_0}{\partial n_x}, \tag{11.39b}$$

$$\frac{\partial\psi_B}{\partial n_y}=-jk\beta(\hat{n}_p\cdot\hat{n}_y^s)\psi_B-\frac{j\gamma}{2\pi}k\pi\sqrt{1-\beta^2}$$
$$H_1^{(2)}\left(kR_0\sqrt{1-\beta^2}\right)e^{jk\beta(x_3+h)}\frac{\partial R_0}{\partial n_y}, \tag{11.39c}$$

$$\frac{\partial^2\psi_B}{\partial n_x\partial n_y}=-\frac{j\gamma}{2\pi}k\pi\sqrt{1-\beta^2}e^{jk\beta(x_3+h)}\Bigg[k\sqrt{1-\beta^2}$$
$$H_0^{(2)}\left(kR_0\sqrt{1-\beta^2}\right)\frac{\partial R_0}{\partial n_x}\frac{\partial R_0}{\partial n_y}+H_1^{(2)}\left(kR_0\sqrt{1-\beta^2}\right)$$
$$\left(\frac{\partial^2 R_0}{\partial n_x\partial n_y}+jk\beta(\hat{n}_p\cdot\hat{n}_x)\frac{\partial R_0}{\partial n_y}-\frac{1}{R_0}\frac{\partial R_0}{\partial n_x}\frac{\partial R_0}{\partial n_y}\right)\Bigg]$$
$$-jk\beta(\hat{n}_p\cdot\hat{n}_y^s)\frac{\partial\psi_B}{\partial n_x}. \tag{11.39d}$$

In Appendix E, all associated expressions to formulas (11.38a)–(11.38d) and (11.39a)–(11.39d) are given.

Numerical stability

The integral I_W and its derivatives are singular for certain values of β and δ for which $\mathrm{Re}(W(t_1))=0$, but this singularity is integrable and the integral can be computed by inserting a semi-circular contour in the upper half of the t-complex plane with radio ε centered on the singular point t_1 similar to the procedure for the Sommerfeld integral (see Fig. 11.8). The integral (11.38a) and its derivatives converge due to the exponential term $e^{-kR^s t}$ except in the case $R^s=0$, which correspond to the case where source and field point coincide on the ground.

11.5.3 Formula with Complex Image Sources

This formula was already presented in Sect. 5.4.1. The correction term can be written as

$$C(\gamma) = \frac{j\gamma}{2\pi} \int_0^{+\infty} \left(\frac{e^{-jk\sqrt{\rho^2+(x_3+h-j\zeta)^2}}}{\sqrt{\rho^2 + (x_3 + h - j\zeta)^2}} \right) e^{j\gamma\zeta} d\zeta, \tag{11.40}$$

with $\rho = \left((x_1 - y_1)^2 + (x_2 - y_2)^2\right)^{1/2}$.

Ochmann showed that (11.40) can be interpreted as a series of monopoles with complex source positions, which can be denoted as "complex monopoles." A complex position $\vec{y}_c$ can be written as

$$\vec{y}_c(\beta) = \vec{y} - j\beta\hat{n}_c, \tag{11.41}$$

where $\vec{y}$ and $\hat{n}_c$ are real vectors and β a real scalar. The complex monopole does not radiate uniformly in space but with a focusing in a direction given by the unit vector $\hat{n}_c$. The degree of focusing is proportional to the product $k\beta$. Formula (11.41) can be rewritten as

$$C(\gamma) = j2\gamma \int_0^{+\infty} \frac{e^{-jkR_c^s(\zeta)}}{4\pi R_c^s(\zeta)} e^{j\gamma\zeta} d\zeta, \quad R_c^s(\zeta) = \left|\vec{x} - \vec{y}_c^s(\zeta)\right|, \tag{11.42a}$$

where $\vec{y}_c^s$ is the mirror image of the complex position $\vec{y}_c$, $\hat{n}_c = (0, 0, 1)$ and $\hat{n}_c^s$ is its mirror image: $\hat{n}_c^s = (0, 0, -1)$. The complex monopoles in (11.42a) are all focused in a direction perpendicular to the plane, pointing not toward the acoustic domain but toward the opposite direction so that their contribution to the acoustic domain has the maximum value at the ground. Hence, $C(\gamma)$ can be seen as the sound radiation due to surface waves on the infinite plane whose amplitude decays exponentially.

The derivatives of (11.42a) are not very difficult to obtain. They are given by:

$$\frac{\partial C(\gamma)}{\partial n_y} = \frac{j\gamma}{2\pi} \int_0^{+\infty} \left(\frac{1}{R_c^s} + jk \right) \frac{\vec{R}_c^s \cdot \hat{n}_y^s}{R_c^s} \frac{e^{-jkR_c^s}}{R_c^s} e^{j\gamma\zeta} d\zeta, \tag{11.42b}$$

$$\frac{\partial C(\gamma)}{\partial n_x} = -\frac{j\gamma}{2\pi} \int_0^{+\infty} \left(\frac{1}{R_c^s} + jk \right) \frac{\vec{R}_c^s \cdot \hat{n}_x}{R_c^s} \frac{e^{-jkR_c^s}}{R_c^s} e^{j\gamma\zeta} d\zeta, \tag{11.42c}$$

$$\frac{\partial^2 C(\gamma)}{\partial n_x \partial n_y} = \frac{j\gamma}{2\pi} \int_0^{+\infty} \left[\left(k^2 - \frac{3jk}{R_c^s} - \frac{3}{R_c^{s2}} \right) \frac{\vec{R}_c^s \cdot \hat{n}_y^s}{R_c^s} \frac{\vec{R}_c^s \cdot \hat{n}_x}{R_c^s} \right.$$

$$+\left(\frac{1}{R_c^{s2}}+\frac{jk}{R_c^s}\right)\hat{n}_y^s\cdot\hat{n}_x\bigg]\frac{\mathrm{e}^{-jkR_c^s}}{R_c^s}\mathrm{e}^{j\gamma\zeta}\mathrm{d}\zeta. \tag{11.42d}$$

Numerical stability

The main advantage of formulas (11.42a)–(11.42d) is that they converge for all types of impedances and are regular except for the case that both, source and field point, lie on the plane. The convergence can be easily demonstrated analyzing the exponential term $\mathrm{e}^{-jkR_c^s}$. The complex distance $R_c^s(\zeta)$ is written as

$$R_c^s(\zeta)=\left|\vec{x}-\vec{y}_c^s(\zeta)\right|=\pm\left(\frac{\sqrt{a^2+b^2}+a}{2}\right)^{1/2}$$
$$\pm\, j\operatorname{sgn}(b)\left(\frac{\sqrt{a^2+b^2}-a}{2}\right)^{1/2},$$

with $a=\rho^2+(x_3+h)^2-\zeta^2$ and $b=-2(x_3+h)\zeta$. We must take the root with positive real part in order to fulfill the radiation condition. With that selection, $\mathrm{Im}(R_c^s(\zeta))<0$ since $b<0$. Therefore, $\mathrm{e}^{-jkR_c^s}$ decays exponentially by increasing ζ, **independently** of the impedance type. If the impedance has a real component ($\mathrm{Re}(Z_0)>0$), the decay is even faster.

The integrand has an oscillating part given by $k\left|\mathrm{Re}(R_c^s)-\mathrm{Im}(Z_0)\zeta/|Z_0|^2\right|$. The oscillation is stronger if $\mathrm{Im}(Z_0)<0$ (spring-like) than if $\mathrm{Im}(Z_0)>0$ (mass-like) so that the convergence of the integral would take more time in the former case.

11.5.4 Approximated Formula

The most used approximation to calculate the sound propagation above an impedance ground is the Weyl-van der Pol formula. It was first deduced for electromagnetic wave theory. In acoustics, the formula was obtained by different authors using different methods [Rud47], [Ing51], [Chi75]. The correction term reads then

$$C=-(1-R_g)g^sF,\quad F=\sqrt{\pi}\sqrt{-\rho}e^{-\rho}\left(1-\mathrm{erf}\left(\sqrt{-\rho}\right)\right), \tag{11.43a}$$

$$R_g=\frac{\delta-\beta}{\delta+\beta},\quad \rho=-\frac{j}{2}kR^s(\delta+\beta)^2,\quad \delta=\frac{z+h}{R^s} \tag{11.44}$$

and is supposed to be valid for $kR^s\gg 1$

The function F can be rewritten using the Faddeeva function w

$$w(z)=\mathrm{e}^{-z^2}(1-\mathrm{erf}(-jz)) \tag{11.45}$$

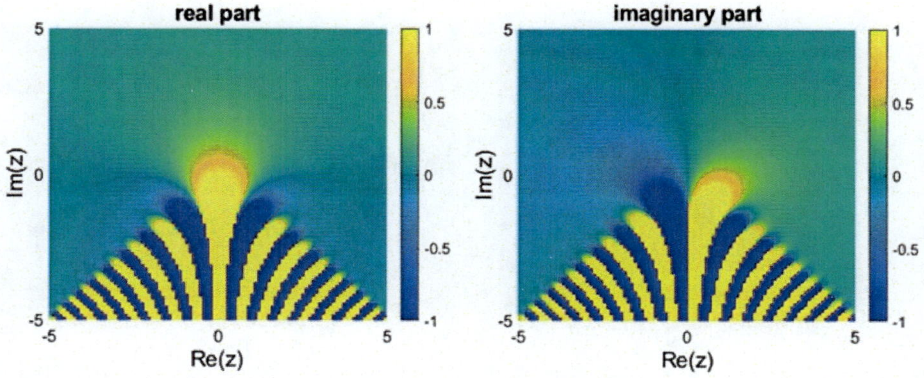

Fig. 11.9 Faddeeva function for complex argument

as

$$F = \sqrt{\pi}\sqrt{-\rho}w\big(j\sqrt{-\rho}\big). \tag{11.46}$$

The square root can be expressed analytically as

$$\varepsilon = \sqrt{-\rho} = \pm\frac{1+j}{2}\sqrt{kR^s}(\delta+\beta), \tag{11.47}$$

where the sign has to be chosen so that $\mathrm{Re}(\varepsilon) > 0$ to ensure a finite value of $w(z)$ since the Faddeeva function $w(z)$ grows asymptotically for $\arg(z) \in [-3\pi/4, -\pi/4]$ as shown in Fig. 11.9.

The derivatives of (11.42a) are given by

$$\frac{\partial C(\gamma)}{\partial n_y} = \frac{\partial R_g}{\partial n_y}g^s F - (1-R_g)\left(\frac{\partial g^s}{\partial n_y}F + \frac{\partial F}{\partial n_y}g^s\right), \tag{11.43b}$$

$$\frac{\partial C(\gamma)}{\partial n_x} = \frac{\partial R_g}{\partial n_x}g^s F - (1-R_g)\left(\frac{\partial g^s}{\partial n_x}F + \frac{\partial F}{\partial n_x}g^s\right), \tag{11.43c}$$

$$\begin{aligned}\frac{\partial^2 C(\gamma)}{\partial n_x \partial n_y} = {} & \frac{\partial^2 R_g}{\partial n_x \partial n_y}g^s F + \frac{\partial R_g}{\partial n_y}\left(\frac{\partial g^s}{\partial n_x}F + g^s\frac{\partial F}{\partial n_x}\right) \\ & + \frac{\partial R_g}{\partial n_x}\left(\frac{\partial g^s}{\partial n_y}F + \frac{\partial F}{\partial n_y}g^s\right) \\ & - (1-R_g)\left(\frac{\partial^2 g^s}{\partial n_x \partial n_y}F + g^s\frac{\partial^2 F}{\partial n_x \partial n_y} + \frac{\partial g^s}{\partial n_y}\frac{\partial F}{\partial n_x} + \frac{\partial g^s}{\partial n_x}\frac{\partial F}{\partial n_y}\right).\end{aligned} \tag{11.43d}$$

In Appendix F, all associated expressions to formulas (11.43a)–(11.43d) are given.

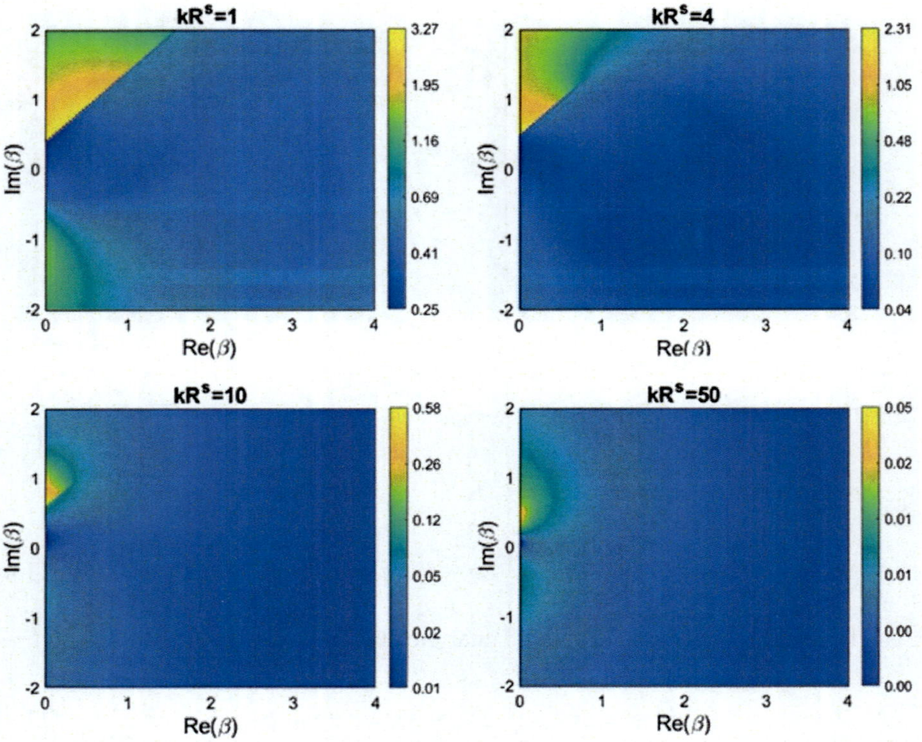

Fig. 11.10 Maximum relative error E_C

Taraldsen [Tar05b] points out that (11.43a) can be easily obtained from the expression of Ingard [Ing51] by neglecting the quadratic term in the square root. But as it has been previously mentioned, the Ingard's formula is incomplete and do not considers the surface waves that appear by grounds with spring-like character. Therefore, this approximated formula will be less accurate for this type of impedances. Figure 11.10 shows the maximum relative error of the correction term and its derivatives E_A

$$E_A = \max\left\{\frac{|C_A - C_M|}{|C_M|}, \frac{|\partial C_A/\partial n_x - \partial C_M/\partial n_x|}{|\partial C_M/\partial n_x|}, \right.$$
$$\left. \frac{\left|\partial C_A/\partial n_y - \partial C_M/\partial n_y\right|}{\left|\partial C_M/\partial n_y\right|}, \frac{\left|\partial^2 C_A/\partial n_x \partial n_y - \partial^2 C_M/\partial n_x \partial n_y\right|}{\left|\partial^2 C_M/\partial n_x \partial n_y\right|}\right\}, \tag{11.48}$$

where the subscripts A and M denote approximation and complex monopoles respectively, for different values of the normalized admittance β at four different values of kR^s and δ for fixed normal vectors $\hat{n}_x$ and $\hat{n}_y$. The colors are scaled logarithmically to appreciate better the differences.

As expected, the relative error decreases with increasing kR^s, which lies in the range [25, 327%] by $kR^s = 1$ to [0–5%] by $kR^s = 50$. One can also observe that the error is a lot bigger for the admittance of spring-like grounds ($\beta > 0$) and

Table 11.2 Relative error of the different formulas

	$kR^s = 1$	$kR^s = 4$	$kR^s = 10$	$kR^s = 50$
δ	0.4	0.5	0.6	0.7
$E_S \times 100\,(\%)$	8.7×10^{-7}	8.9×10^{-4}	1.1×10^{-7}	6.2×10^{-6}
$E_T \times 100\,(\%)$	3.7×10^{-6}	1.0×10^{-6}	7.1×10^{-5}	1.2×10^{-5}
$E_A \times 100\,(\%)$	327	231.3	58.4	4.5

in the region of $\text{Im}(\beta) > \text{Re}(\beta) + \delta$, since the approximation does not include the surface waves that appear for the admittances in that region. In Table 11.2, we resume the values of the error including also the errors of the Sommerfeld and Thomasson formulas defined similarly to 11.48.

11.5.5 Comparison of the Performance of the Half-Space Formulas in a BEM Formulation

To study the performance of the four mentioned formulas in a BEM formulation, we will perform a monopole test for the half-space problem. It is usually used to test the accuracy of any BEM implementation or to determine the magnitude of the discretization error. It consists in the following procedure, given the discretized model of a structure, the normal velocity on the surface is assumed to be equal to the normal particle velocity produced by a monopole located inside it. The pressure on the surface and subsequently other acoustic quantities like sound power or pressure at certain field points are then obtained by means of a BEM calculation. Since the exact values are known, the accuracy of the simulated results can be determined.

For the structure, we consider a small sphere with radius $R = 0.15$ m centered at $(0, 0, 0.15)$ and the discretized model has 296 quadrangular elements. The size of the elements ensures to have at least six elements per wavelength up to about 2000 Hz. We take a total of 360 field points distributed on two parallel lines at $x \in [0.16, 31]$, $(y, z) = (0, 0.15)$ and $(y, z) = (0, 0)$.

Before we study the performance of the half-space formulas, we perform first a monopole test in free space to determine the magnitude of the discretization error. The monopole is located at the position $(0.03, 0.02, 0.17)$, i.e., not in the center of sphere to avoid having a symmetric velocity distribution. Two calculations are made, one without the Burton and Miller (B&M) method and the other with it. Figure 11.11 shows the geometry of the calculation and the errors in dB in the sound power and in the sound pressure and normal velocity at the field points with and without B&M.

We can clearly see the presence of two critical frequencies when the B&M method is not included. When using the B&M, the error in the sound power is bigger away from the critical frequencies, but near them it is smaller and in all frequencies remain below 0.5 dB. At the field points, the errors are even smaller than in the sound power. Since critical frequencies are not known in a general complex structure, it is always better to include the B&M method. In the present case, however, if we select the

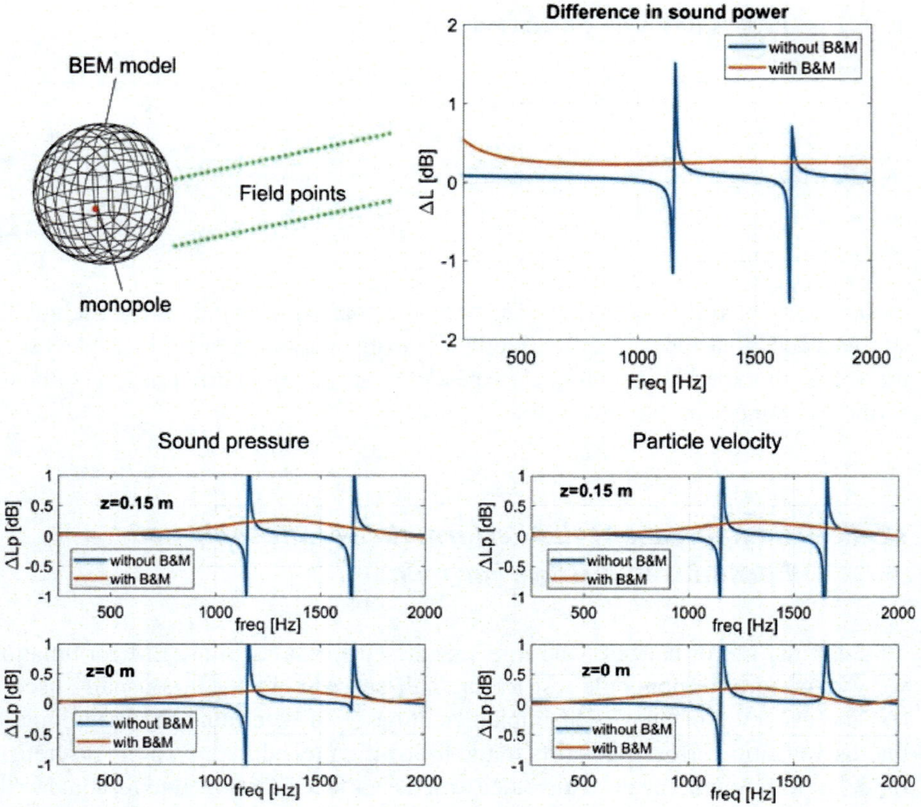

Fig. 11.11 Results of the monopole test in free space

frequency 500 Hz, we will obtain very accurate results if we do not consider the B&M method since the first critical frequency lies above 1000 Hz and the discretization error is minimum.

Knowing that the discretization error is negligible, we perform the monopole test in the half-space. We consider the three types of impedances with $|Z_0| = 0.8$: mass ($Z_0 = j|Z_0|$), absorbing ($Z_0 = |Z_0|$) and spring ($Z_0 = -j|Z_0|$) character.

Inserting the ground at $z = 0$, we will have the sphere sitting on the ground and the first row of field points also on the ground. Now, the known analytical formula for the monopole will be given by (11.28) with $C(\gamma)$ defined as (11.31), (11.37), (11.40), or (11.43) for the Sommerfeld, Thomasson, complex image sources, or van der Pol formulas, respectively.

As it was discussed in the previous section, the error of the van der Pol formula is larger for the case the ground has a spring-character, and if the monopole is located near the ground, we find that the sound pressure obtained with the van der Pol formula at the field points lying on the ground is strongly underestimated. By shifting the sphere 1 m above the ground, the sound pressure at the ground is very close to the values of the other formulas. Figure 11.12 show the sound pressure level computed

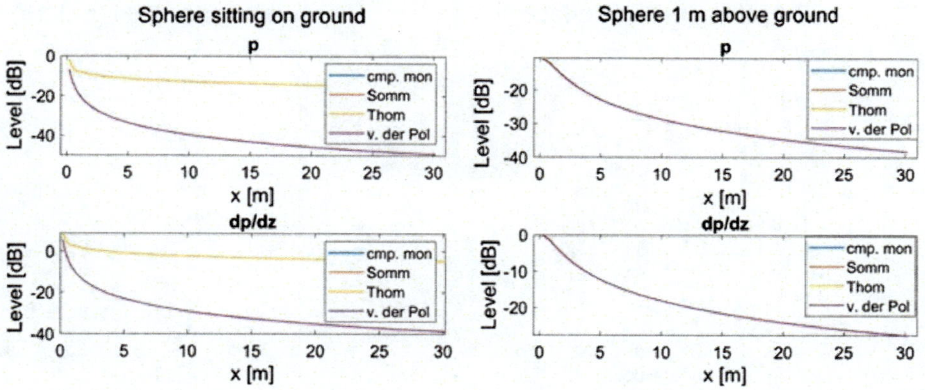

Fig. 11.12 Comparison of sound pressure levels and their normal derivatives on the ground for a monopole above the spring-like ground

by the four formulas at the field points on the ground when the sphere is sitting on the ground (left) and when the sphere is 1 m above the ground (right). In the first case, the first three formulas provide the same results while the fourth formula differs. In the second case, all four formulas provide very close results. So, for the test we consider the sphere 1 m above the ground.

Now that we know that the discretization error is negligible and all formulas provide the same or very close results for the monopole above the ground (but not too close to it), we perform the monopole test for the three types of impedance with $|Z_0| = 0.8$. The normal velocity on the surface of the sphere is given and the sound pressure is computed. From both pressure and velocity on the surface, pressure and velocity at the field points are calculated. The results from the BEM simulation can be compared with the analytical results and the accuracy can be obtained. Figures 11.13, 11.14 and 11.15 show the error in the pressure on the surface (upper left), at the field points (upper right and lower left) and in the impedance (lower right). Even though there are results for four formulas, in all cases one sees two different curves. That happens because the first three formulas provide practically the same results while the other curve belongs to the fourth formula (van der Pol).

The magnitude of the error in the pressure on the surface is very small, below 0.02 dB and larger at the field points. For the field points at $z = 1.15$, the error varies near the sphere but becomes approximately constant far away from it. For the field points on the ground, the error tends to increase linearly with the distance to the sphere, but up to 30 m away it remains below 0.7 dB. Since pressure and velocity are known at the field points on the ground, the impedance can be determined and it should be equal to the predetermined Z_0. We find that the error in the impedance is small up to 30 m away and also tends to grow linearly with the distance to the sphere.

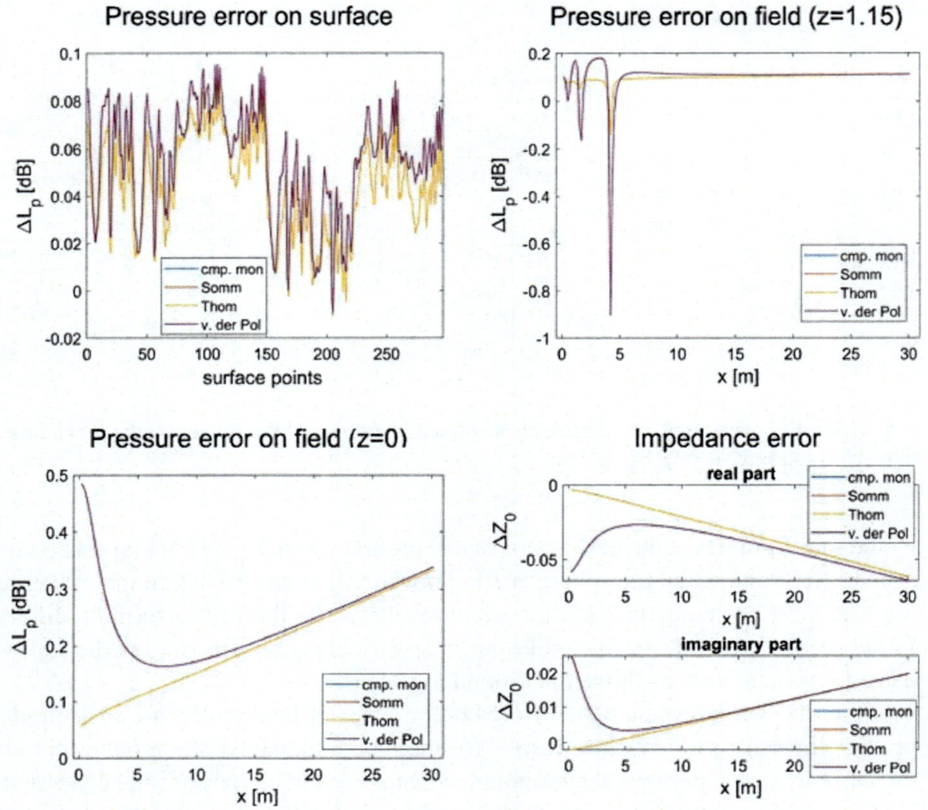

Fig. 11.13 Error in the sound pressure and in the impedance for a mass-like ground

The monopole test shows that from the four studied formulas, the three exact ones provide the same results and the approximated one produce a bigger error but the accuracy is still acceptable.

The time of computation differs between the methods. For a comparison, the ratio of times, with respect to the time of computation of the formula with complex monopoles, is considered. Table 11.3 summarizes the results.

11.6 Example of Application: Acoustic Thin Barrier Above Impedance Ground

Acoustic barriers are solid constructions that are built between noise sources (often traffic noise) and populated areas. Their function is to reduce the level of sound at the receiving side, since sound diffraction makes a complete blocking of sound impossible. The simplest form of a barrier is a vertical rigid wall of height H and thickness D. The presence of the barrier produces a bright and a shadow zone, being the latter

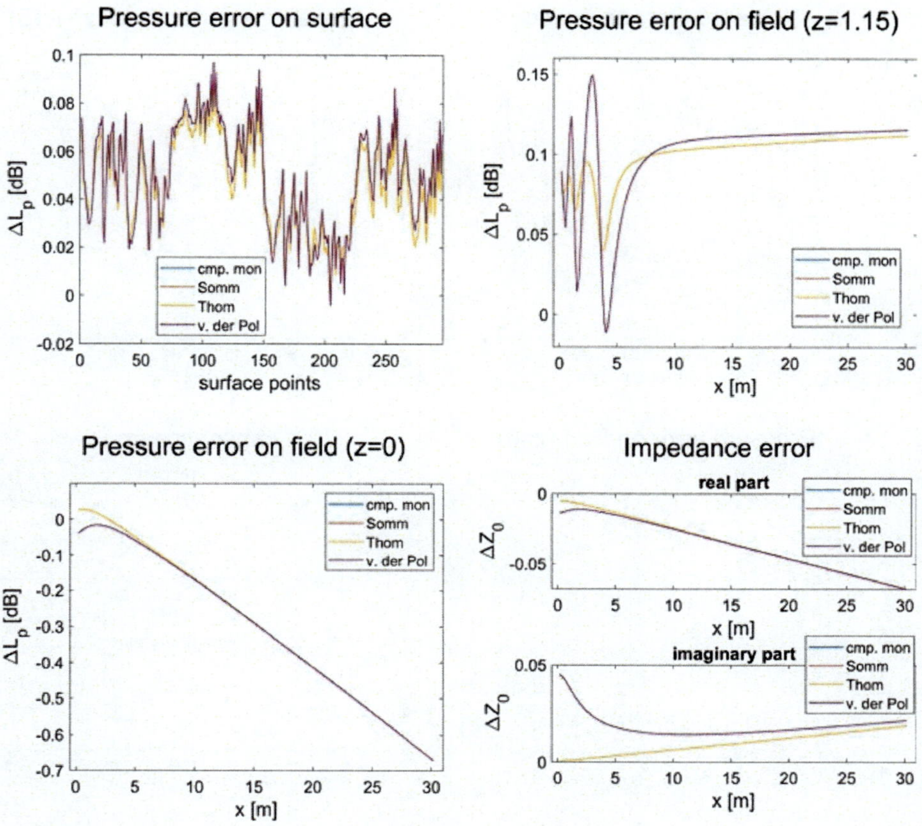

Fig. 11.14 Error in the sound pressure and in the impedance for an absorbing ground

the zone where there is no free path between source and receiver. Figure 11.16 shows how the wave fronts go around the edge of the barrier.

The effectiveness of the acoustic barrier is evaluated with the *insertion loss* (*IL*), which is given by the expression

$$\mathrm{IL} = 10\log_{10}\left(\frac{|p_{\mathrm{nb}}|^2}{|p_{\mathrm{wb}}|^2}\right), \tag{11.49}$$

where p_{nb} is the pressure without barrier and p_{wb} is the pressure with the barrier.

11.6.1 Numerical Model

In order to obtain accurate results, the number of elements per wavelength should be at least 6. For thin barriers and at low frequencies, $\lambda/6$ can be larger than the thickness D of the barrier. In that case, numerical problems may arise due to ill-conditioned

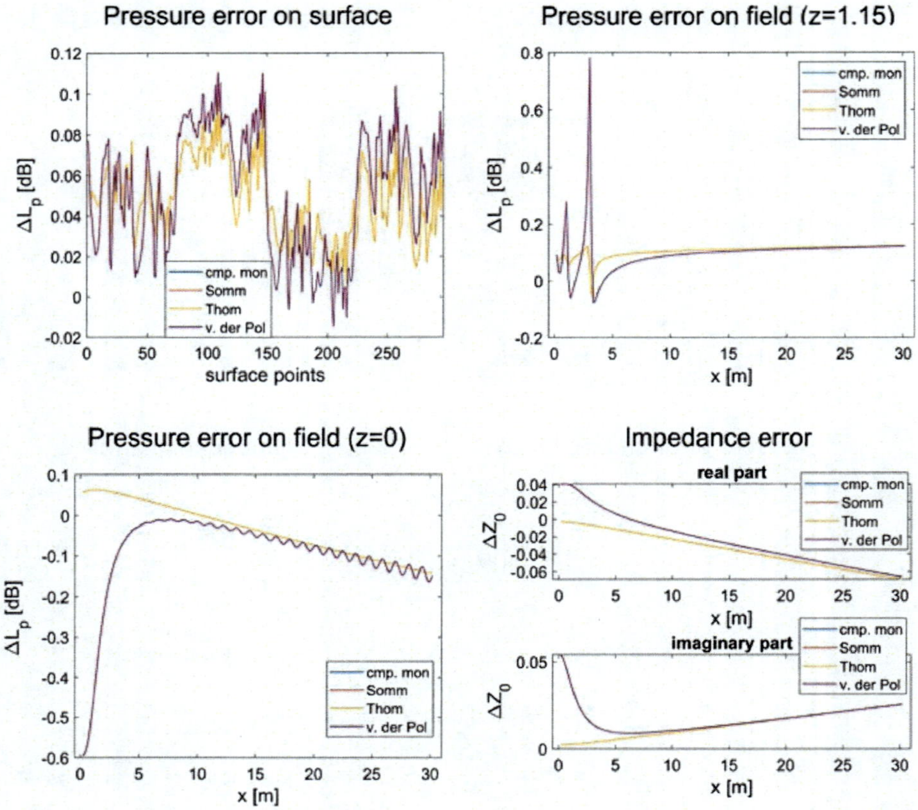

Fig. 11.15 Error in the sound pressure and in the impedance for a spring-like ground

Table 11.3 Relative time of calculation for the monopole test

Impedance type	Sommerfeld	Thomasson	Van der Pol
Mass-like	1.92	0.89	0.0015
Absorbing	2.00	0.95	0.0024
Spring-like	2.86	0.85	0.0014

matrices or almost singular integrals (narrow gap problem [Vla00]). To solve the problems and ensure good results, the size of the element should be taken as $\Delta x = \min(\lambda/6, D)$ which increases the number of elements. Therefore, the use of the direct BEM can be very expensive in time of computation and main memory, especially if the barrier is long. In that case, it is more efficient to consider an infinitely thin barrier and to apply an indirect BEM approach.

Fig. 11.16 Sound waves in front and behind a noise barrier

Figure 11.17 shows the two meshes for finitely thin and infinitely thin barriers, respectively. The mesh on the right has more than 6 times more elements than the one on the left.

In [Pis16], a comparison between the values of pressure level obtained with a finitely thin barrier with the results obtained with an infinitely thin barrier is presented and the differences remain below 1 dB.

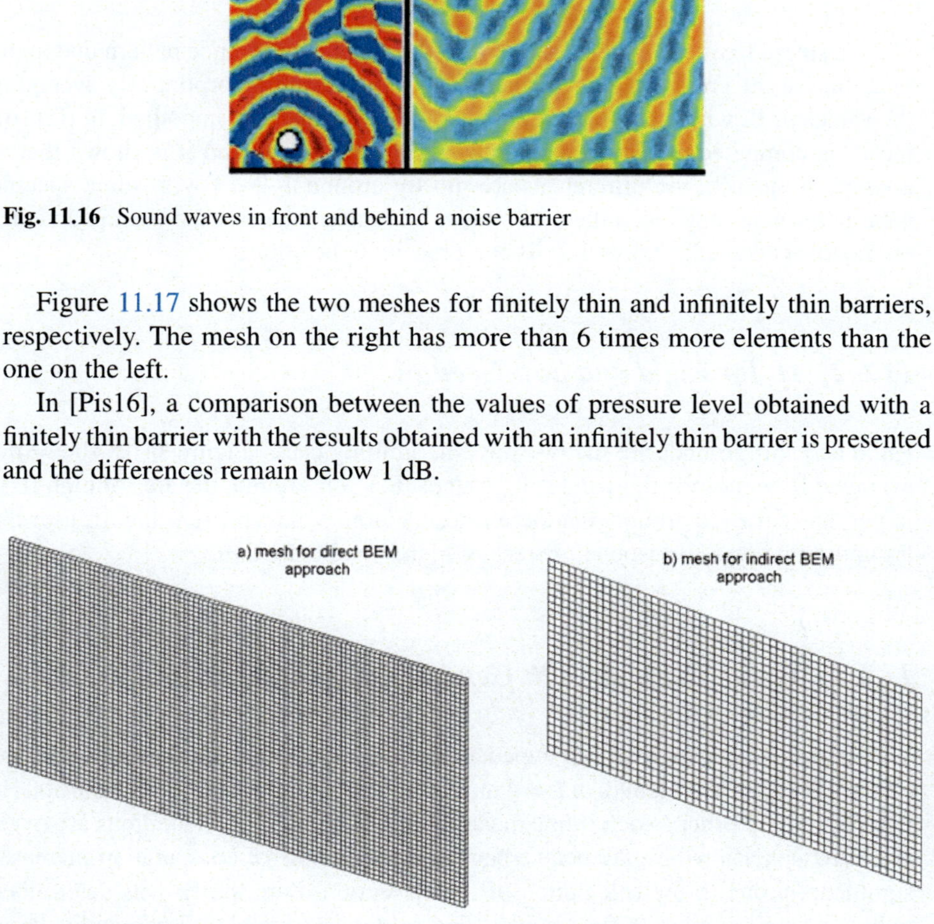

Fig. 11.17 Meshes for the direct and indirect BEM approaches

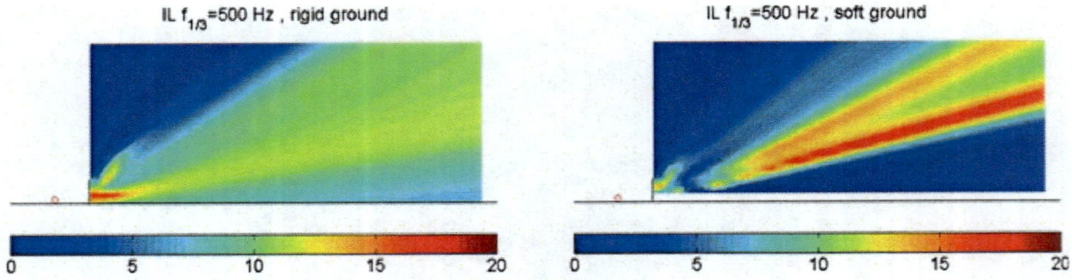

Fig. 11.18 Comparison of IL for the 3D thin barrier

At a single frequency, the IL shows a well-defined interference pattern due to the reflection of the ground specially when the ground has no absorption. By averaging the values of IL within a third octave band, the pattern is a bit smoothed. In [Pis16], the IL is computed at a series of points behind the barrier and it is shown that at a single frequency, the differences are mostly around 1–2 dB with some specific locations where they reach the 6 dB value. The differences for the averaged values on the third octave lie below 1.2 dB at all receiving positions.

11.6.2 IL for Rigid and Soft Ground

Rigid and soft grounds are the two possible limiting cases. Figure 11.18 shows the averaged IL values for rigid and soft grounds. For soft ground, the area of high IL is larger than for rigid ground, but there is also a large zone with very low IL near the ground. Red areas correspond to areas of high sound reduction.

11.6.3 IL for Grounds with General Type of Impedance

The IL of a thin barrier above an impedance ground was investigated. The barrier has a height $H = 2$ m and a length of $L = 4$ m. The simulations are performed at the 500 Hz octave band. In order to determine the averaged IL value, the calculations are made at 11 frequencies within the octave band. Since the time of computation increases significantly due to the calculation of the correction term, the IL was determined only at a small number of field points. The field points are placed perpendicular to the barrier at a distance from the barrier varying from 0.2 to 15.2 m and at a height of 1.5 m. The calculations are performed using the formula with complex monopoles and consider the three types of impedances: mass-like, absorbing, and spring-like.

The curves of the averaged IL for the different values of impedances are plotted together with the two limiting cases, namely, soft ground and rigid ground. Figures 11.19, 11.20 and 11.21 show the results for an absorbing impedance, mass-like impedance and spring-like impedance respectively.

According to the results, the major improvement is obtained by an absorbing impedance followed by the mass-like impedance. The spring-like impedance on the contrary cannot provide any advantage compared to the rigid and soft cases. More details can be found in [Pis16].

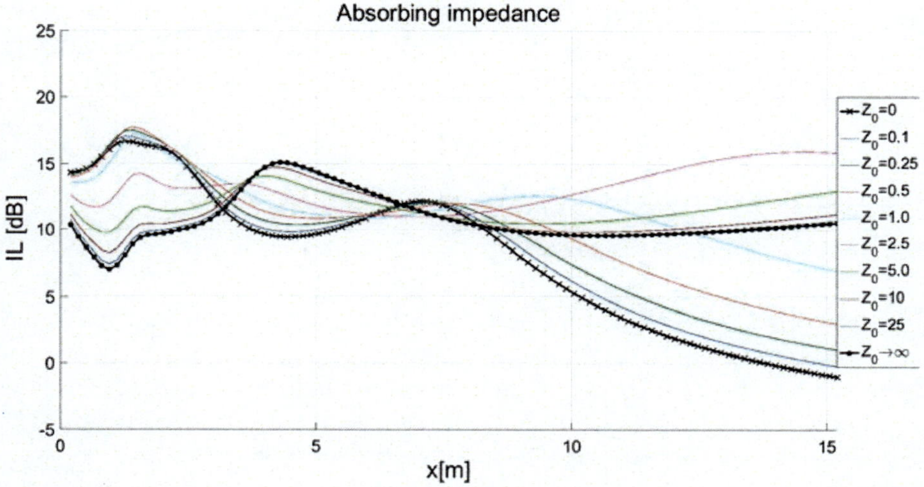

Fig. 11.19 IL of a barrier on absorbing ground

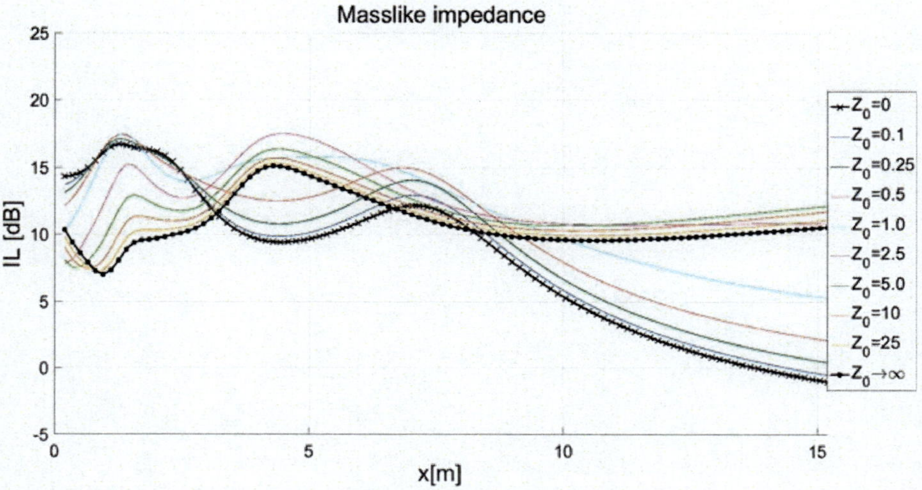

Fig. 11.20 IL of a barrier on a mass-like ground

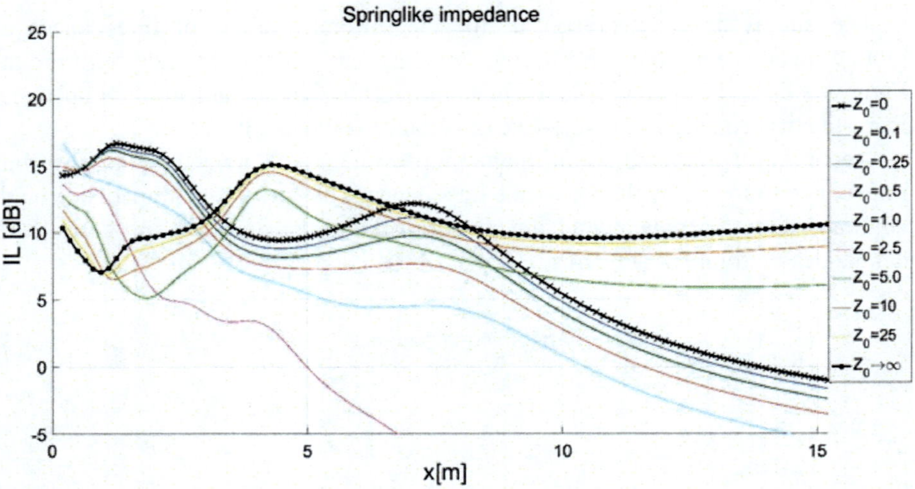

Fig. 11.21 IL of a barrier on a spring-like ground

Chapter 12
Final Remarks and Outlook

Throughout the book, we have studied stationary complex sources and their application only in the frequency domain. Time domain solutions for point sources and line sources over mass-like and absorbing planes can be found by using the theory of real image sources and inverse Fourier transform [Och11, Och18, Och20]. However, complex sources in time domain require a more demanding theory as in frequency domain, "*where operators are elliptic and life is simple.*" (quoted from [Kai03, p. R294] Furthermore, not only complex sources can be investigated in the time domain, but time itself can also be complexified.

In one of the first works on complex sources, Felsen already mentioned the possibility of allowing both the spatial and temporal coordinates of a source to accept complex values (see [Fel76a], p. 40). Tygel and Hubral give an excellent overview of transient wave propagation phenomena in [Tyg87]. In Chaps. 13 and 14 of [Tyg87], the authors consider not only complex source point coordinates, but also complex times and establish a connection with analytical signals. The theory is applied to problems in geophysics. Heyman and Felsen [Hey89, Hey01] investigate so-called *complex-source pulsed beam fields.* They show that the complexification of time variables leads to a number of novel phenomena such as "*short-pulse wave packets, pulsed beams, complex-source pulsed beams, pulsed-beam-excited propagation and diffraction*" etc. (see [Hey01], p. 1592). In [Hey00] this work is transferred from the electromagnetic to the acoustic field, and energy expressions such as the time-dependent intensity of pulsed sound beams are studied. In the overview article [Kai03], Kaiser provides extensive material on the fundamentals of physical wavelets in complex spacetime, which is presented from the perspective of the mathematical physicist and has not yet been fully utilized for acoustic science.

Since we have only considered stationary complex sources, moving sources are not treated in the book. This topic is particularly important from a practical point of view. The theory of moving sources contains elements from the frequency and from the time domain as shown in [Och13a], where a Lorentz transformation is used to obtain a wave equation with a stationary monopole source. Ziolkowski [Zio85]

M. Ochmann and R. Piscoya, *Theory and Application of Acoustic Sources Using Complex Analysis*, https://doi.org/10.1007/978-981-33-6040-2_12

considers moving, complex sources, which he calls modulated, moving Gaussian pulses representing exact solutions of the time-dependent wave equation.

Unfortunately, because of their complexity, these two advanced topics cannot be treated in the present book. But we hope that we can cover these exciting subjects in a future publication.

Appendix A
Different Expressions for the Driving Source of a Complex 3D Monopole

We compare our result of Theorem 3

$$\tilde{\delta}_3(x, y, z) = M_3(r, a)\delta(z) + i D_3(r, a)\delta'(z) \tag{4.51}$$

with

$$\begin{aligned} D_3(r, a) &= \frac{1}{2\pi} \int_0^\infty \frac{1}{\kappa} \sinh(\kappa a) J_0(k_r r) k_r \mathrm{d}k_r \\ &= \frac{H(a-r)}{2\pi} \frac{\cosh\left(k\sqrt{a^2 - r^2}\right)}{\sqrt{a^2 - r^2}} \end{aligned} \tag{4.52}$$

and

$$\begin{aligned} M_3 &= \frac{1}{2\pi} \int_0^\infty \cosh(\kappa a) J_0(k_r r) k_r \mathrm{d}k_r \\ &= \frac{1}{2\pi} \left\{ \frac{\delta(r-a)}{r} \cosh(ka) \right. \\ &\quad \left. - aH(a-r) \left[\frac{\cosh\left(k\sqrt{a^2 - r^2}\right)}{\left(a^2 - r^2\right)^{3/2}} - k \frac{\sinh\left(k\sqrt{a^2 - r^2}\right)}{a^2 - r^2} \right] [((\cdot)) - (a)] \right\} \\ &= R_1 + R_2 + R_3 \\ &\text{(the definition of the three summands is obvious)} \end{aligned} \tag{4.53}$$

with the expression F of Tagirdzhanov et al. (see [Tag15], p. 319 which is a corrected version of the source case as given by the same authors in [Tag11])

M. Ochmann and R. Piscoya, *Theory and Application of Acoustic Sources Using Complex Analysis*, https://doi.org/10.1007/978-981-33-6040-2

$$\delta_{\mathrm{Tar}} = F/(-4\pi) = M_{\mathrm{tar}}(r,a)\delta(z) + i D_{\mathrm{tar}}(r,a)\delta'(z) \tag{A.1}$$

where

$$\begin{aligned} M_{\mathrm{tar}}(r,a) &= \frac{1}{2\pi}\delta(r-a) + \frac{ka}{2\pi}H(r-a)\frac{\sinh\left(k\sqrt{a^2-r^2}\right)}{\left(a^2-r^2\right)} \\ &- \frac{a}{2\pi}H(r-a)\frac{1}{\left(a^2-r^2\right)^{\frac{3}{2}}}\left[\cosh\left(k\sqrt{a^2-r^2}\right)(\cdot) - (a)\right] \end{aligned} \tag{A.2}$$

and

$$D_{\mathrm{tar}}(r,a) = -\frac{a}{i}\frac{H(r-a)}{2\pi}\frac{\cosh\left(k\sqrt{a^2-r^2}\right)}{\sqrt{a^2-r^2}} \tag{A.3}$$

with $r = \sqrt{x^2+y^2}$ and $\kappa = \sqrt{k^2-k_r^2}$.

Note that we have to divide F by (-4π) due to the definitions (1) and (4) used in [Tag15], in order to make it comparable with Eqs. (4.48) and (4.49). The dipole term D_{tar} of (A.1) is identical with the dipole term (4.52) when a is replaced by i which also gives the correct dimension! The different minus signs are caused by the fact that we use $z + ia$ instead of $z - ia$ [see the reasoning directly after Eq. (4.47)].

However, at first sight, the monopole terms look very different. It is obvious that the factor $1/r$ is missing in the first delta function of (A.2) which is clear only from dimensional reasons. Thus, we have

$$\begin{aligned} M_{\mathrm{tar}}^{\mathrm{corr}}(r,a) &= \frac{1}{r}\frac{1}{2\pi}\delta(r-a) + \frac{ka}{2\pi}H(r-a)\frac{\sinh\left(k\sqrt{a^2-r^2}\right)}{\left(a^2-r^2\right)} \\ &- \frac{a}{2\pi}H(r-a)\frac{1}{\left(a^2-r^2\right)^{\frac{3}{2}}}\left[\cosh\left(k\sqrt{a^2-r^2}\right)(\cdot) - (a)\right] \\ &=: T_1 + T_2 + T_3 \end{aligned} \tag{A.4}$$

consisting of three obvious summands.

Now, we consider the difference $M_3(r,a) - M_{\mathrm{tar}}^{\mathrm{corr}}(r,a)$ acting on a sufficiently smooth test function f:

$$D = \int_0^\infty \left[M_3(r,a) - M_{\mathrm{tar}}^{\mathrm{corr}}(r,a)\right] f(r) r \mathrm{d}r = D_1 + D_{23}. \tag{A.5}$$

The difference between the first summands gives

$$\begin{aligned} D_1 &= \int_0^\infty [R_1 - T_1] f(r) r \mathrm{d}r \\ &= \frac{1}{2\pi} \int_0^\infty [\cosh(ka) - 1] \delta(r-a) f(r) \mathrm{d}r \\ &= \frac{1}{2\pi} [\cosh(ka) - 1] f(a). \end{aligned} \tag{A.6}$$

A simple calculation shows that the difference between the last two summands gives

$$\begin{aligned} D_{23} &= \int_0^\infty [R_2 + R_3 - T_2 - T_3] f(r) r \mathrm{d}r \\ &= \frac{a}{2\pi} \int_0^a \left[\frac{\cosh\left(k\sqrt{a^2 - r^2}\right) f(a) - f(a)}{\left(a^2 - r^2\right)^{\frac{3}{2}}} - \frac{k \sinh\left(k\sqrt{a^2 - r^2}\right) f(a)}{\left(a^2 - r^2\right)} \right] r \mathrm{d}r. \end{aligned} \tag{A.7}$$

We show below that

$$D_{23} = \frac{1}{2\pi} [1 - \cosh(ka)] f(a) = -D_1, \tag{A.8}$$

and thus the total difference $D = D_1 + D_{23}$ is zero. In conclusion, the monopole terms are equal in the distributional sense, too, and thus the driving sources in 3D given by Theorem 3 and by the formula found in [Tag15] are identical, if the two small corrections mentioned above are taken into account.

The only thing missing is the proof of Eq. (A.8). For that reason, we consider [cp. (A.7)]

$$I := \int_0^a \left[\frac{\cosh\left(k\sqrt{a^2 - r^2}\right) - 1}{\left(a^2 - r^2\right)^{\frac{3}{2}}} - \frac{k \sinh\left(k\sqrt{a^2 - r^2}\right)}{\left(a^2 - r^2\right)} \right] r \mathrm{d}r. \tag{A.9}$$

By using the substitution $r = a \sin t$, we get

$$I := \int_0^{\pi/2} \left[\frac{\cosh(ka \cos t) - 1}{(a \cos t)^3} - \frac{k \sinh(ka \cos t)}{(a \cos t)^2} \right] a \sin t a \cos t \mathrm{d}t \tag{A.10}$$

or

$$I := k \int_0^{\pi/2} \left[\frac{\cosh(ka \cos t) - 1}{(ka \cos t)^2} - \frac{\sinh(ka \cos t)}{ka \cos t} \right] ka \sin t \, \mathrm{d}t. \tag{A.11}$$

Since

$$\left(\frac{\cosh(ka \cos t)}{(ka \cos t)^2} - \frac{\sinh(ka \cos t)}{ka \cos t} \right) ka \sin t = \frac{\mathrm{d}}{\mathrm{d}t} \left(\frac{\cosh(ka \cos t)}{ka \cos t} \right),$$

the integral becomes

$$\begin{aligned}
I &:= k \int_0^{\pi/2} \left[\frac{\mathrm{d}}{\mathrm{d}t} \left(\frac{\cosh(ka \cos t)}{ka \cos t} \right) - \frac{ka \sin t}{(ka \cos t)^2} \right] \mathrm{d}t \\
&= k \int_0^{\pi/2} \frac{\mathrm{d}}{\mathrm{d}t} \left[\frac{\cosh(ka \cos t)}{ka \cos t} - \frac{1}{ka} \left(\frac{1}{\cos t} \right) \right] \mathrm{d}t \\
&= k \left[\frac{\cosh(ka \cos t)}{ka \cos t} - \frac{1}{ka} \left(\frac{1}{\cos t} \right) \right]_0^{\pi/2} \\
&= \left[\frac{\cosh(ka \cos t) - 1}{a \cos t} \right]_0^{\pi/2} \\
&= -\frac{1}{a} [\cosh(ka) - 1]
\end{aligned} \tag{A.12}$$

Here, the upper limit is zero

$$\lim_{t \to \pi/2} \left[\frac{\sinh(ka \cos t) ka \sin t}{a \sin t} \right] = 0,$$

where we have used L'Hospital's rule. Finally, we obtain

$$D_{23} = \frac{a}{2\pi} f(a) I = -\frac{1}{2\pi} [\cosh(ka) - 1] f(a) = -D_1.$$

which completes the proof of Eq. (A.8).

Appendix B
Proof of the Expression for the Driving Sources of the Half-Plane Problem

We proof formula (5.65):

$$\chi_2^{(2)} := -D\sigma_2^{(2)} = 2i\gamma \int_0^\infty \mathrm{e}^{i\gamma\tau}\tilde{\delta}_2(x, z_h; \tau)\mathrm{d}\tau$$

$$= -\frac{\mathrm{e}^{ik\sqrt{1-\frac{1}{Z_0^2}}|x|}}{Z_0\sqrt{1-1/Z_0^2}}\left\{\frac{ik}{Z_0}\delta(z_h) - \delta'(z_h)\right\} \tag{5.65}$$

with

$$\tilde{\delta}_2(x, z_h; \tau) = M_2(\tau)\delta(z_h) + iD_2(\tau)\delta'(z_h) \tag{5.64}$$

and [see Eqs. (4.23) and (4.24)]

$$M_2(\tau) = \frac{1}{2\pi}\int_{-\infty}^{\infty} \cosh\left(\sqrt{k^2 - k_x^2}\tau\right)\mathrm{e}^{ik_x x}\mathrm{d}k_x,$$

$$D_2(\tau) = \frac{1}{2\pi}\int_{-\infty}^{\infty} \frac{\sinh\left(\sqrt{k^2 - k_x^2}\tau\right)}{\sqrt{k^2 - k_x^2}}\mathrm{e}^{ik_x x}\mathrm{d}k_x. \tag{B.1}$$

Thus, we get immediately

$$\chi_2^{(2)} = \frac{i\gamma}{\pi}\int_0^\infty \mathrm{e}^{i\gamma\tau}\left\{\delta(z_h)\int_{-\infty}^{\infty} \cosh\left(\sqrt{k^2 - k_x^2}\tau\right)\mathrm{e}^{ik_x x}\mathrm{d}k_x\right.$$

M. Ochmann and R. Piscoya, *Theory and Application of Acoustic Sources Using Complex Analysis*, https://doi.org/10.1007/978-981-33-6040-2

$$+i\delta'(z_h)\int_{-\infty}^{\infty}\frac{\sinh\left(\sqrt{k^2-k_x^2}\tau\right)}{\sqrt{k^2-k_x^2}}\mathrm{e}^{ik_x x}\mathrm{d}k_x\Bigg\}\mathrm{d}\tau. \tag{B.2}$$

Assuming that we can change the order of integration, we get

$$\begin{aligned}\chi_2^{(2)} &= \frac{i\gamma}{\pi}\int_{-\infty}^{\infty}\Bigg\{\delta(z_h)\int_0^{\infty}\mathrm{e}^{i\gamma\tau}\cosh\left(\sqrt{k^2-k_x^2}\tau\right)\mathrm{d}\tau\\ &\quad+i\delta'(z_h)\int_0^{\infty}\mathrm{e}^{i\gamma\tau}\frac{\sinh\left(\sqrt{k^2-k_x^2}\tau\right)}{\sqrt{k^2-k_x^2}}\mathrm{d}\tau\Bigg\}\mathrm{e}^{ik_x x}\mathrm{d}k_x\\ &=: \frac{i\gamma}{\pi}\int_{-\infty}^{\infty}\left\{A\delta_h(z)+iB\delta'(z_h)\right\}\mathrm{e}^{ik_x x}\mathrm{d}k_x\\ &= 2i\gamma\left[F_x^{-1}\{A\}\delta(z_h)+F_x^{-1}\{B\}\delta'(z_h)\right]\end{aligned} \tag{B.3}$$

The integrals A and B are known from Eqs. (5.40) and (5.41)

$$A=\int_0^{\infty}\cosh(\kappa_2\tau)\mathrm{e}^{i\gamma\tau}\mathrm{d}\tau=\frac{i\gamma}{\gamma^2+\kappa_2^2}=\frac{k}{Z_0}\frac{-1}{(k\mu)^2-k_x^2} \tag{B.4}$$

and

$$B=\frac{i}{\kappa}\int_0^{\infty}\sinh(\kappa_2\tau)\mathrm{e}^{i\gamma\tau}\mathrm{d}\tau=\frac{-i}{\gamma^2+\kappa_2^2}=\frac{-i}{(k\mu)^2-k_x^2} \tag{B.5}$$

under the condition of convergence as described in Eq. (5.42). Here again, we have used

$$\mu=\sqrt{1-\frac{1}{Z_0^2}},\ \gamma=\frac{ik}{Z_0},\ \text{and}\ \kappa_2=\sqrt{k^2-k_x^2}.$$

We remark that Eqs. (B.3), (B.4), and (B.5) lead to the useful formula

$$\begin{aligned}\chi_2^{(2)} &= 2i\gamma F_x^{-1}\left[A\delta(z_h)+B\delta'(z_h)\right]\\ &= -2\gamma F_x^{-1}\left[\frac{1}{\gamma^2+\kappa_2^2}\left(\gamma\delta(z_h)-\delta'(z_h)\right)\right].\end{aligned} \tag{B.5a}$$

The inverse Fourier transforms of A and B can be seen from Eqs. (6.14)/(6.15)

$$
\begin{aligned}
F_x^{-1}\{A\} &= \frac{k}{2\pi Z_0}\int_{-\infty}^{\infty}\frac{-1}{(k\mu)^2-k_x^2}\mathrm{e}^{ik_x x}\mathrm{d}k_x \\
&= \frac{k}{Z_0}\frac{i}{2k\mu}\mathrm{e}^{ik\mu|x|} \\
&= \frac{i}{2Z_0\sqrt{1-1/Z_0^2}}\mathrm{e}^{ik\mu|x|},
\end{aligned}
\tag{B.6}
$$

$$
\begin{aligned}
F_x^{-1}\{B\} &= \frac{i}{2\pi}\int_{-\infty}^{\infty}\frac{-1}{(k\mu)^2-k_x^2}\mathrm{e}^{ik_x x}\mathrm{d}k_x \\
&= \frac{-1}{2k\mu}\mathrm{e}^{ik\mu|x|} \\
&= \frac{-1}{2k\sqrt{1-1/Z_0^2}}\mathrm{e}^{ik\mu|x|},
\end{aligned}
\tag{B.7}
$$

where we have again defined the square root function $\mu = \sqrt{1-\frac{1}{Z_0^2}}$ such that $\operatorname{Im}\mu > 0$.

By inserting Eqs. (B.6) and (B.7) into Eq. (B.3), we get

$$
\begin{aligned}
\chi_2^{(2)} &= 2i\gamma\left[F_x^{-1}\{A\}\delta(z_h) + F_x^{-1}\{B\}\delta'(z_h)\right] \\
&= \frac{-k}{Z_0}\frac{1}{\sqrt{1-1/Z_0^2}}\left[\frac{i}{Z_0}\delta(z_h) - \frac{1}{k}\delta'(z_h)\right]\mathrm{e}^{ik\sqrt{1-1/Z_0^2}|x|} \\
&= \frac{-1}{Z_0\sqrt{1-1/Z_0^2}}\left[\frac{ik}{Z_0}\delta(z_h) - \delta'(z_h)\right]\mathrm{e}^{ik\sqrt{1-1/Z_0^2}|x|}
\end{aligned}
\tag{B.8}
$$

which is the RHS of (5.65), and thus, the proof is complete.

Appendix C
Proof of the Expression for the Driving Sources of the Half-Space Problem

We proof formula (5.76):

$$\chi_3^{(2)} := -D\sigma_3^{(2)} = 2i\gamma \int_0^\infty e^{i\gamma\tau} \tilde{\delta}_3(x, z_h; \tau) d\tau$$
$$= \frac{-k}{2Z_0} H_0^{(1)}\left(kr\sqrt{1 - \frac{1}{Z_0^2}}\right)\left\{\frac{ik}{Z_0}\delta(z_h) - \delta'(z_h)\right\}. \quad (5.76)$$

with

$$\tilde{\delta}_3(x, y, z_h; \tau) = M_3(\tau)\delta(z_h) + iD_3(\tau)\delta'(z_h), r = \sqrt{x^2 - y^2}, \quad (5.75)$$

and [see Eqs. (4.52) and (4.53)]

$$M_3(\tau) = \frac{1}{2\pi} \int_0^\infty \cosh\left(\sqrt{k^2 - k_r^2}\tau\right) J_0(k_r r) k_r dk_r,$$
$$D_3(\tau) = \frac{1}{2\pi} \int_0^\infty \frac{\sinh\left(\sqrt{k^2 - k_r^2}\tau\right)}{\sqrt{k^2 - k_r^2}} J_0(k_r r) k_r dk_r. \quad (C.1)$$

Putting the last three equations together gives

$$\chi_3^{(2)} = \frac{i\gamma}{\pi} \int_0^\infty e^{i\gamma\tau} \left\{ \delta(z_h) \int_0^\infty \cosh\left(\sqrt{k^2 - k_r^2}\tau\right) J_0(k_r r) k_r dk_r \right.$$
$$\left. + i\delta(z_h) \int_0^\infty \frac{\sinh\left(\sqrt{k^2 - k_x^2}\tau\right)}{\sqrt{k^2 - k_r^2}} J_0(k_r r) k_r dk_r \right\} d\tau. \quad (C.2)$$

M. Ochmann and R. Piscoya, *Theory and Application of Acoustic Sources Using Complex Analysis*, https://doi.org/10.1007/978-981-33-6040-2

Assuming that we can change the order of integration, we obtain

$$\chi_3^{(2)} = \frac{i\gamma}{\pi}\int_0^\infty \Bigg\{ \delta(z_h)\int_0^\infty \mathrm{e}^{i\gamma\tau}\cosh\left(\sqrt{k^2-k_r^2}\tau\right)\mathrm{d}\tau$$

$$+i\delta'(z_h)\int_0^\infty \mathrm{e}^{i\gamma\tau}\frac{\sinh\left(\sqrt{k^2-k_x^2}\tau\right)}{\sqrt{k^2-k_r^2}}\mathrm{d}\tau\Bigg\} J_0(k_r r)k_r\mathrm{d}k_r$$

$$=: \frac{i\gamma}{\pi}\int_0^\infty \left\{A\delta(z_h)+iB\delta'(z_h)\right\}J_0(k_r r)k_r\mathrm{d}k_r$$

$$= \frac{i\gamma}{\pi}\left[H_r^{-1}\{A\}\delta(z_h)+H_r^{-1}\{B\}\delta'(z_h)\right]. \tag{C.3}$$

Again, the integrals A and B are known from Eqs. (5.40) and (5.41), [see also (B.4) and (B.5)]

$$A = \int_0^\infty \cosh(\kappa_3\tau)\mathrm{e}^{i\gamma\tau}\mathrm{d}\tau = \frac{i\gamma}{\gamma^2+\kappa_3^2} = \frac{k}{Z_0}\frac{-1}{(k\mu)^2-k_r^2} \tag{C.4}$$

and

$$B = \frac{i}{\kappa_3}\int_0^\infty \sinh(\kappa_3\tau)\mathrm{e}^{i\gamma\tau}\mathrm{d}\tau = \frac{-i}{\gamma^2+\kappa_3^2} = \frac{-i}{(k\mu)^2-k_r^2} \tag{C.5}$$

under the condition of convergence as given in Eq. (5.42). Here again, we have used

$$\mu = \sqrt{1-\frac{1}{Z_0^2}},\ \gamma = \frac{ik}{Z_0},\ \text{and}\ \kappa_3 = \sqrt{k^2-k_r^2}.$$

We remark that Eqs. (C.3), (C.4), and (C.5) lead to the useful formula

$$\chi_3^{(2)} = \frac{i\gamma}{\pi}H_r^{-1}\left[A\delta(z_h)+B\delta'(z_h)\right]$$

$$= -\frac{1}{2\pi}H_r^{-1}\left[\frac{2\gamma}{\gamma^2+\kappa_3^2}\left(\gamma\delta(z_h)-\delta'(z_h)\right)\right]. \tag{C.5a}$$

The inverse Hankel transforms of A and B can be seen from Eq. (6.34)

$$H_r^{-1}\{A\} = \frac{k}{Z_0}\int_0^\infty \frac{-1}{(k\mu)^2-k_r^2}J_0(k_r r)k_r\mathrm{d}k_r$$

$$= 2\pi \frac{k}{Z_0} \frac{i}{4} H_0^{(1)}(k\mu r), \tag{C.6}$$

$$H_r^{-1}\{B\} = \int_{-\infty}^{\infty} \frac{-i}{(k\mu)^2 - k_r^2} J_0(k_r r) k_r \mathrm{d}k_r$$

$$= 2\pi i \frac{i}{4} H_0^{(1)}(k\mu r). \tag{C.7}$$

By inserting Eqs. (C.6) and (C.7) into Eq. (C.3), we get

$$\begin{aligned} \chi_3^{(2)} &= \frac{i\gamma}{\pi}\left[H_r^{-1}\{A\}\delta(z_h) + H_r^{-1}\{B\}\delta'(z_h)\right] \\ &= \frac{-2k}{Z_0}\left[\frac{k}{Z_0}\delta(z_h) + i\delta'(z_h)\right]\frac{i}{4}H_0^{(1)}(k\mu r) \\ &= \frac{-k}{2Z_0}\left[\frac{ik}{Z_0}\delta(z_h) - \delta'(z_h)\right]H_0^{(1)}(k\mu r) \end{aligned} \tag{C.8}$$

which is the RHS of (5.76), and thus, the proof is complete.

Appendix D
Exact Image Theory

Reflection Coefficient

$$\hat{R}_w(p) = \frac{\alpha p - \sqrt{p^2+1}}{\alpha p + \sqrt{p^2+1}} \cdot \frac{\alpha p - \sqrt{p^2+1}}{\alpha p - \sqrt{p^2+1}} = \frac{\left(\alpha p - \sqrt{p^2+1}\right)^2}{(\alpha^2-1)p^2-1} \tag{D.1}$$

Continuing the analysis of the numerator

$$\left(\alpha p - \sqrt{p^2+1}\right)^2 = \alpha^2 p^2 + 1 + p^2 - 2\alpha p\sqrt{p^2+1}$$

we add and subtract $\alpha(2p^2+1)$ and rearrange terms

$$\begin{aligned}\left(\alpha p - \sqrt{p^2+1}\right)^2 &= \alpha^2 p^2 + 1 + p^2 + \alpha\left(2p^2 + 1 - 2p\sqrt{p^2+1} - 2p^2 - 1\right)\\ &= \alpha^2 p^2 + 1 + p^2 - \alpha(2p^2+1) + \alpha\left(p - \sqrt{p^2+1}\right)^2\\ &= p^2(\alpha-1)^2 - (\alpha-1) + \alpha\left(p - \sqrt{p^2+1}\right)^2\end{aligned}$$

then multiply and divide the first two terms by $\alpha + 1$

$$= \frac{\alpha-1}{\alpha+1}\left(p^2(\alpha^2-1) - 1 - \alpha\right) + \alpha\left(p - \sqrt{p^2+1}\right)^2 \tag{D.2}$$

The reflection coefficient becomes

$$\hat{R}_w(p) = \frac{\left(\alpha p - \sqrt{p^2+1}\right)^2}{(\alpha^2-1)p^2-1} = \frac{\frac{\alpha-1}{\alpha+1}\left(p^2(\alpha^2-1) - 1 - \alpha\right) + \alpha\left(p - \sqrt{p^2+1}\right)^2}{(\alpha^2-1)p^2-1}$$

M. Ochmann and R. Piscoya, *Theory and Application of Acoustic Sources Using Complex Analysis*, https://doi.org/10.1007/978-981-33-6040-2

By performing the change of variable $s = q - u$ in I_1 and I_2, we find that $I_1 = I_3$ and $I_2 = I_4$.

I_3 can be written as

$$I_3 = \frac{e^{\gamma q}}{2}\left(\int_0^{+\infty} e^{-\gamma u}\frac{J_2(u)}{u}du - \underbrace{\int_q^{+\infty} e^{-\gamma u}\frac{J_2(u)}{u}du}_{I_5}\right) \tag{D.6}$$

By performing the change of variable $s = u - q$, I_5 can be written as

$$I_5 = e^{-\gamma q}\int_0^{+\infty} e^{-\gamma s}\frac{J_2(s)}{s}ds$$

and I_3 becomes

$$I_3 = \frac{e^{\gamma q}}{2}\int_0^{+\infty} e^{-\gamma u}\frac{J_2(u)}{u}du - \frac{1}{2}\int_0^{+\infty} e^{-\gamma u}\frac{J_2(u+q)}{u+q}du \tag{D.7}$$

In a similar way, I_4 can be written as

$$I_4 = \int_0^{+\infty} \frac{e^{-\gamma u}}{2}\frac{J_2(q-u)}{q-u}du - \underbrace{\int_q^{+\infty} \frac{e^{-\gamma u}}{2}\frac{J_2(q-u)}{q-u}du}_{I_6} \tag{D.8}$$

By performing the change of variable $s = u - q$, I_6 can be written as

$$I_6 = -\frac{e^{-\gamma q}}{2}\int_0^{+\infty} e^{-\gamma s}\frac{J_2(s)}{s}ds$$

where the relation $J_\nu(ze^{jm\pi}) = e^{jm\nu\pi}J_\nu(z)$ was used (i.e., $J_2(-z) = J_2(z)$) and I_4 becomes

$$I_4 = \frac{1}{2}\int_0^{+\infty} e^{-\gamma u}\frac{J_2(q-u)}{q-u}du + \frac{e^{-\gamma q}}{2}\int_0^{+\infty} e^{-\gamma u}\frac{J_2(u)}{u}du \tag{D.9}$$

Combining (D.7) and (D.9), we obtain for I

$$
\begin{aligned}
I &= 2(I_3 - I_4) \\
&= 2\left(\frac{e^{\gamma q}}{2}\int_0^{+\infty} e^{-\gamma u}\frac{J_2(u)}{u}du - \frac{1}{2}\int_0^{+\infty} e^{-\gamma u}\frac{J_2(u+q)}{u+q}du\right. \\
&\quad \left. -\frac{1}{2}\int_0^{+\infty} e^{-\gamma u}\frac{J_2(q-u)}{q-u}du - \frac{e^{-\gamma q}}{2}\int_0^{+\infty} e^{-\gamma u}\frac{J_2(u)}{u}du\right) \\
&= 2\sinh(\gamma q)\underbrace{\int_0^{+\infty} e^{-\gamma u}\frac{J_2(u)}{u}du}_{I_7} - \int_0^{+\infty} e^{-\gamma u}\left(\frac{J_2(u+q)}{u+q} + \frac{J_2(q-u)}{q-u}\right)du
\end{aligned}
$$

I_7 has the form of a Laplace transform and we can find in a table that

$$
I_7 = \frac{1}{2}\left(\sqrt{\gamma^2+1} - \gamma\right)^2 = \frac{1}{2}\left(\frac{\sqrt{\gamma^2+1}-\gamma}{\sqrt{\gamma^2+1}+\gamma}\right) \tag{D.10}
$$

Inserting $\gamma^2 = 1/(\alpha^2 - 1)$ in (D.10)

$$
I_7 = \frac{1}{2}\left(\frac{\alpha-1}{\alpha+1}\right) \tag{D.11}
$$

and I is therefore

$$
I = \left(\frac{\alpha-1}{\alpha+1}\right)\sinh(\gamma q) - \left(\int_0^{+\infty} e^{-\gamma u}\frac{J_2(u+q)}{u+q}du + \int_0^{+\infty} e^{-\gamma u}\frac{J_2(q-u)}{q-u}du\right) \tag{D.12}
$$

Inserting (D.12) into (D.4), we finally obtain for the source density

$$
r(q) = -\alpha\gamma \int_0^{+\infty} e^{-\gamma u}\left(\frac{J_2(u+q)}{u+q} + \frac{J_2(q-u)}{q-u}\right)du \tag{D.13}
$$

Transmission Coefficient $\hat{T}_w$

The transmission coefficient $\hat{T}_w$ is found to be

$$\hat{T}_w = \frac{2k_{2z}k_{1z}}{\alpha k_{1z} + k_{2z}} \frac{e^{jk_{1z}h}}{k_{1z}} \tag{D.14}$$

By introducing the variable $Q = k_{1z} + k_{2z}$ and considering $k_{1z} = \sqrt{k_{2z}^2 - B^2}$, one can write k_{1z} and k_{2z} in terms of Q and B.

$$k_{1z} = \frac{Q^2 + B^2}{2Q} \text{ and } k_{2z} = \frac{Q^2 - B^2}{2Q} \tag{D.15}$$

Then

$$\begin{aligned} \frac{k_{1z}}{\alpha k_{1z} + k_{2z}} &= \frac{Q^2 - B^2}{(\alpha + 1)Q^2 - (\alpha - 1)B^2} \\ &= \frac{1}{\alpha + 1}\left(\frac{\left(\frac{B}{Q}\right)^2 - 1}{\left(\frac{\alpha-1}{\alpha+1}\right)\left(\frac{B}{Q}\right)^2 - 1}\right) \end{aligned}$$

By defining $y = \left(\frac{B}{Q}\right)^2$ and $a = \frac{\alpha-1}{\alpha+1}$

$$\frac{k_{1z}}{\alpha k_{1z} + k_{2z}} = \frac{1}{\alpha + 1}\left(\frac{1 - y}{1 - ay}\right) = \frac{1}{\alpha + 1}\left(1 + \frac{a - 1}{a}\frac{ay}{1 - ay}\right) \tag{D.16}$$

Considering the Taylor expansion

$$\frac{1}{1 - x} = \sum_{n=0}^{\infty} x^n. \tag{D.17}$$

(D.16) becomes

$$\frac{k_{1z}}{\alpha k_{1z} + k_{2z}} = \frac{1}{\alpha + 1}\left(1 + \frac{a - 1}{a}\sum_{n=1}^{\infty}(ay)^n\right)$$

and since $\frac{a-1}{a} = -\frac{2}{\alpha-1}$

$$\frac{k_{1z}}{\alpha k_{1z} + k_{2z}} = \frac{1}{\alpha + 1} - \frac{2}{\alpha^2 - 1}\sum_{n=1}^{\infty}\left(\frac{\alpha - 1}{\alpha + 1}\right)^n\left(\frac{B}{k_{1z} + k_{2z}}\right)^{2n} \tag{D.18}$$

Finally, the transmission coefficient $\hat{T}_w$ is given by

$$\hat{T}_w = k_{2z}\left(\frac{2}{\alpha+1} - \frac{4}{\alpha^2-1}\sum_{n=1}^{\infty}\left(\frac{\alpha-1}{\alpha+1}\right)^n\left(\frac{B}{k_{1z}+k_{2z}}\right)^{2n}\right)\frac{e^{jk_{1z}}}{k_{1z}} \tag{D.19}$$

Derivative of $F(\alpha, B, q)$

The function $F(\alpha, B, q)$ is given by

$$F(\alpha, B, q) = \frac{2}{\alpha+1}J_0(Bq) - \frac{4}{\alpha^2-1}\sum_{n=1}^{\infty}\underbrace{\left(\frac{\alpha-1}{\alpha+1}\cdot\frac{h-H_T(q)}{h+H_T(q)}\right)^n J_{2n}(Bq)}_{I_8} \tag{D.20}$$

We compute first the derivative of I_8

$$\frac{dI_8}{dq} = n\left(\frac{\alpha-1}{\alpha+1}\cdot\frac{h-H_T}{h+H_T}\right)^{n-1}\left(\frac{\alpha-1}{\alpha+1}\right)\underbrace{\left(\frac{-H_T'}{h+H_T} - \frac{-(h-H_T)H_T'}{(h+H_T)^2}\right)}_{I_9}J_{2n}(Bq)$$
$$+\left(\frac{\alpha-1}{\alpha+1}\cdot\frac{h-H_T}{h+H_T}\right)^n BJ_{2n}'(Bq) \tag{D.21}$$

Since $H_T = \sqrt{q^2+h^2} \rightarrow H_T' = q/H_T$ and I_9 is given by

$$I_9 = \frac{-2hH_T'}{(h+H_T)^2} = \frac{-2hq}{H_T(h+H_T)^2}$$

Multiplying and dividing by $h - H_T$, we obtain

$$\begin{aligned} I_9 &= \frac{-2hq}{H_T}\left(\frac{h-H_T}{h+H_T}\right)\frac{1}{(h+H_T)(h-H_T)} \\ &= \frac{-2hq}{H_T}\left(\frac{h-H_T}{h+H_T}\right)\frac{1}{h^2-H_T^2} \\ &= \frac{2hB}{H_TBq}\left(\frac{h-H_T}{h+H_T}\right) \end{aligned} \tag{D.22}$$

Inserting (D.22) into (D.21), we have

$$\frac{\mathrm{d}I_8}{\mathrm{d}q} = B\left(\frac{\alpha-1}{\alpha+1}\cdot\frac{h-H_T}{h+H_T}\right)^n\left(\frac{h}{H_T}\frac{2n}{Bq}J_{2n}(Bq) + J'_{2n}(Bq)\right) \tag{D.23}$$

We introduce into (D.23) the relation

$$J'_{2n}(Bq) = J_{2n-1}(Bq) - \frac{2n}{Bq}J_{2n}(Bq)$$

and obtain

$$\frac{\mathrm{d}I_8}{\mathrm{d}q} = B\left(\frac{\alpha-1}{\alpha+1}\cdot\frac{h-H_T}{h+H_T}\right)^n\left(\left(\frac{h}{H_T}-1\right)\frac{2n}{Bq}J_{2n}(Bq) + J_{2n-1}(Bq)\right) \tag{D.24}$$

Then, we use the relation

$$\frac{2n}{Bq}J_{2n}(Bq) = \frac{J_{2n-1}(Bq) + J_{2n+1}(Bq)}{2}$$

and

$$\frac{\mathrm{d}I_8}{\mathrm{d}q} = \frac{B}{2}\left(\frac{\alpha-1}{\alpha+1}\cdot\frac{h-H_T}{h+H_T}\right)^n\left(\left(\frac{h}{H_T}-1\right)J_{2n+1}(Bq) + \left(\frac{h}{H_T}+1\right)J_{2n-1}(Bq)\right) \tag{D.25}$$

The derivative of $F(\alpha, B, q)$ is then

$$\begin{aligned}\frac{\mathrm{d}F(\alpha,B,q)}{\mathrm{d}q} &= -\frac{2B}{\alpha+1}J_1(Bq)\\ &-\frac{2B}{\alpha^2-1}\sum_{n=1}^{\infty}\left(\frac{\alpha-1}{\alpha+1}\cdot\frac{h-H_T}{h+H_T}\right)^n\left(\frac{h}{H_T}+1\right)J_{2n-1}(Bq)\\ &-\frac{2B}{\alpha^2-1}\sum_{n=1}^{\infty}\left(\frac{\alpha-1}{\alpha+1}\cdot\frac{h-H_T}{h+H_T}\right)^n\left(\frac{h}{H_T}-1\right)J_{2n+1}(Bq)\end{aligned} \tag{D.26}$$

We take apart the first term of the first sum and have

$$\begin{aligned}\frac{\mathrm{d}F(\alpha,B,q)}{\mathrm{d}q} &= -\frac{2B}{\alpha+1}J_1(Bq)\left(1+\frac{1}{\alpha+1}\left(\frac{h}{H_T}-1\right)\right)\\ &-\frac{2B}{\alpha^2-1}\sum_{n=2}^{\infty}\left(\frac{\alpha-1}{\alpha+1}\cdot\frac{h-H_T}{h+H_T}\right)^n\left(\frac{h}{H_T}+1\right)J_{2n-1}(Bq)\\ &-\frac{2B}{\alpha^2-1}\sum_{n=1}^{\infty}\left(\frac{\alpha-1}{\alpha+1}\cdot\frac{h-H_T}{h+H_T}\right)^n\left(\frac{h}{H_T}-1\right)J_{2n+1}(Bq)\end{aligned} \tag{D.27}$$

Then make the change $n \to n-1$ in the second sum and obtain

$$\frac{\mathrm{d}F(\alpha,B,q)}{\mathrm{d}q} = -\frac{2B}{(\alpha+1)^2}J_1(Bq)\left(\alpha+\frac{h}{H_T}\right) - \frac{2B}{\alpha^2-1}\sum_{n=2}^{\infty}\left(\frac{\alpha-1}{\alpha+1}\cdot\frac{h-H_T}{h+H_T}\right)^n\left(\frac{h}{H_T}+1\right)J_{2n-1}(Bq) - \frac{2B}{\alpha^2-1}\sum_{n=2}^{\infty}\left(\frac{\alpha-1}{\alpha+1}\cdot\frac{h-H_T}{h+H_T}\right)^{n-1}\left(\frac{h}{H_T}-1\right)J_{2n-1}(Bq) \tag{D.28}$$

Regrouping the last two terms

$$\frac{\mathrm{d}F(\alpha,B,q)}{\mathrm{d}q} = -\frac{2B}{(\alpha+1)^2}J_1(Bq)\left(\alpha+\frac{h}{H_T}\right) - \frac{2B}{\alpha^2-1}\sum_{n=2}^{\infty}\left(\frac{\alpha-1}{\alpha+1}\cdot\frac{h-H_T}{h+H_T}\right)^{n-1} \left(\frac{h}{H_T}-1\right)\left(\frac{\alpha-1}{\alpha+1}+1\right)J_{2n-1}(Bq) \tag{D.29}$$

and finally we get

$$\frac{\mathrm{d}F(\alpha,B,q)}{\mathrm{d}q} = -\frac{2B}{(\alpha+1)^2}J_1(Bq)\left(\alpha+\frac{h}{H_T}\right) - \frac{2B}{(\alpha+1)^2}\left(\frac{2}{\alpha+1}\right)\left(\frac{(h-H_T)^2}{H_T(h+H_T)}\right) \sum_{n=2}^{\infty}\left(\frac{\alpha-1}{\alpha+1}\cdot\frac{h-H_T}{h+H_T}\right)^{n-2} J_{2n-1}(Bq) \tag{D.30}$$

Derivative of $\tilde{F}(\alpha, B, q)$

Since $\tilde{F}(\alpha, B, q)$ has practically the same form as $F(\alpha, B, q)$, its derivative follows the same previous analysis and is written as

$$\frac{\mathrm{d}\tilde{F}(\alpha,B,q)}{\mathrm{d}q} = -\frac{2}{(\alpha+1)^2}J_1(q)\left(\alpha+\frac{h}{\tilde{H}_T}\right)$$

$$-\frac{2}{(\alpha+1)^2}\left(\frac{2}{\alpha+1}\right)\left(\frac{\left(h-\tilde{H}_T\right)^2}{\tilde{H}_T\left(h+\tilde{H}_T\right)}\right)$$

$$\sum_{n=2}^{\infty}\left(\frac{\alpha-1}{\alpha+1}\cdot\frac{h-\tilde{H}_T}{h+\tilde{H}_T}\right)^{n-2} J_{2n-1}(Bq) \qquad \text{(D.31)}$$

Appendix E
Formula of Thomasson

Derivatives of R^s

$$R^s = \sqrt{(\vec{x} - \vec{y}^s) \cdot (\vec{x} - \vec{y}^s)} = \sqrt{(\vec{x}^s - \vec{y}) \cdot (\vec{x}^s - \vec{y})} \tag{E.1}$$

The derivative in the direction of $\hat{n}_y$ is

$$\begin{aligned}\frac{\partial R^s}{\partial n_y} &= \nabla_y\left(R^s\right) \cdot \hat{n}_y \\ &= \left(\frac{\nabla_y((\vec{x}^s - \vec{y}) \cdot (\vec{x}^s - \vec{y}))}{2R^s}\right) \cdot \hat{n}_y \\ &= \frac{-(\vec{x}^s - \vec{y}) \cdot \hat{n}_y}{R^s}\end{aligned}$$

Since $(\vec{x}^s - \vec{y}) \cdot \hat{n}_y = (\vec{x} - \vec{y}^s) \cdot \hat{n}_y^s$

$$\frac{\partial R^s}{\partial n_y} = \frac{-(\vec{x} - \vec{y}^s) \cdot \hat{n}_y{}^s}{R^s} = -\frac{\vec{R}^s \cdot \hat{n}_y^s}{R^s} \tag{E.2}$$

Similarly in the direction of $\hat{n}_x$

$$\frac{\partial R^s}{\partial n_x} = \frac{(\vec{x} - \vec{y}^s) \cdot \hat{n}_x}{R^s} = \frac{\vec{R}^s \cdot \hat{n}_x}{R^s} \tag{E.3}$$

and the second derivative is

$$\frac{\partial^2 R^s}{\partial n_x \partial n_y} = -\nabla_x\left(\frac{\vec{R}^s \cdot \hat{n}_y^s}{R^s}\right) \cdot \hat{n}_x = -\left(\frac{\hat{n}_y^s}{R^s} - \frac{\vec{R}^s \cdot \hat{n}_y^s}{(R^s)^2}\nabla_y R^s\right) \cdot \hat{n}_x$$

M. Ochmann and R. Piscoya, *Theory and Application of Acoustic Sources Using Complex Analysis*, https://doi.org/10.1007/978-981-33-6040-2

Taking into account (E.2) and (E.3)

$$\frac{\partial^2 R^s}{\partial n_x \partial n_y} = -\frac{1}{R^s}\left(\hat{n}_x \cdot \hat{n}^s_y + \frac{\partial R^s}{\partial n_x}\frac{\partial R^s}{\partial n_y}\right) \tag{E.4}$$

Derivatives of δ

$$\delta = \frac{z+h}{R^s} = \frac{(\vec{x}-\vec{y}^s)\cdot\hat{n}_p}{R^s} = \frac{(\vec{x}^s-\vec{y})\cdot\hat{n}^s_p}{R^s} \tag{E.5}$$

The derivative in the direction of $\hat{n}_y$ is

$$\begin{aligned}
\frac{\partial\delta}{\partial n_y} &= \nabla_y\delta\cdot\hat{n}_y \\
&= \nabla_y\left(\frac{(\vec{x}^s-\vec{y})\cdot\hat{n}^s_p}{R^s}\right)\cdot\hat{n}_y \\
&= \left(\frac{-\hat{n}^s_p}{R^s} - \frac{(\vec{x}^s-\vec{y})\cdot\hat{n}^s_p}{(R^s)^2}\nabla_y R^s\right)\cdot\hat{n}_y
\end{aligned}$$

$$\frac{\partial\delta}{\partial n_y} = \frac{-\hat{n}^s_p\cdot\hat{n}_y}{R^s} - \frac{\delta}{R^s}\frac{\partial R^s}{\partial n_y}$$

and since $\hat{n}^s_p\cdot\hat{n}_y = \hat{n}_p\cdot\hat{n}^s_y$

$$\frac{\partial\delta}{\partial n_y} = -\frac{\hat{n}_p\cdot\hat{n}^s_y}{R^s} - \frac{\delta}{R^s}\frac{\partial R^s}{\partial n_y} \tag{E.6}$$

Similarly in the direction of $\hat{n}_x$

$$\begin{aligned}
\frac{\partial\delta}{\partial n_x} &= \nabla_x\delta\cdot\hat{n}_x \\
&= \nabla_x\left(\frac{(\vec{x}-\vec{y}^s)\cdot\hat{n}_p}{R^s}\right)\cdot\hat{n}_x \\
&= \left(\frac{\hat{n}_p}{R^s} - \frac{(\vec{x}-\vec{y}^s)\cdot\hat{n}_p}{(R^s)^2}\nabla_x R^s\right)\cdot\hat{n}_x
\end{aligned}$$

$$\frac{\partial\delta}{\partial n_x} = \frac{\hat{n}_p\cdot\hat{n}_x}{R^s} - \frac{\delta}{R^s}\frac{\partial R^s}{\partial n_x} \tag{E.7}$$

The second derivative is given by

$$\frac{\partial^2\delta}{\partial n_x\partial n_y} = \nabla_x\left(\frac{\partial\delta}{\partial n_y}\right)\cdot\hat{n}_x$$
$$= \nabla_x\left(\frac{-\hat{n}_p\cdot\hat{n}_y^s}{R^s} - \frac{\delta}{R^s}\frac{\partial R^s}{\partial n_y}\right)\cdot\hat{n}_x$$
$$= \frac{\hat{n}_p\cdot\hat{n}_y^s}{(R^s)^2}\frac{\partial R^s}{\partial n_x} - \frac{1}{R^s}\frac{\partial\delta}{\partial n_x}\frac{\partial R^s}{\partial n_y} + \frac{\delta}{(R^s)^2}\frac{\partial R^s}{\partial n_x}\frac{\partial R^s}{\partial n_y} - \frac{\delta}{R^s}\frac{\partial^2 R^s}{\partial n_x\partial n_y} \tag{E.8}$$

By inserting (E.7), we obtain

$$\frac{\partial^2\delta}{\partial n_x\partial n_y} = \frac{1}{(R^s)^2}\left(\delta\hat{n}_x\cdot\hat{n}_y^s + \hat{n}_p\cdot\hat{n}_y^s\frac{\partial R^s}{\partial n_x} - \hat{n}_p\cdot\hat{n}_x\frac{\partial R^s}{\partial n_y} + 3\delta\frac{\partial R^s}{\partial n_x}\frac{\partial R^s}{\partial n_y}\right) \tag{E.9}$$

Derivatives of W

$$W = (\beta+\delta)^2 - j2t(1+\beta\delta) - t^2 \tag{E.10}$$

$$\frac{\partial W}{\partial n_y} = 2(\delta + (1-jt)\beta)\frac{\partial\delta}{\partial n_y} \tag{E.11}$$

$$\frac{\partial W}{\partial n_x} = 2(\delta + (1-jt)\beta)\frac{\partial\delta}{\partial n_x} \tag{E.12}$$

$$\frac{\partial^2 W}{\partial n_x\partial n_y} = 2\left\{\frac{\partial\delta}{\partial n_x}\frac{\partial\delta}{\partial n_y} + (\delta + (1-jt)\beta)\frac{\partial^2\delta}{\partial n_x\partial n_y}\right\} \tag{E.13}$$

Derivatives of I_W

$$I_W = \int_0^\infty \frac{e^{-kR^s t}}{\sqrt{W}}dt \tag{E.14}$$

$$\frac{\partial I_W}{\partial n_x} = -\int_0^\infty \frac{e^{-kR^s t}}{\sqrt{W}}\left(\frac{1}{2W}\frac{\partial W}{\partial n_x} + kt\frac{\partial R^s}{\partial n_x}\right)dt \tag{E.15}$$

$$\frac{\partial I_W}{\partial n_y} = -\int_0^\infty \frac{e^{-kR^s t}}{\sqrt{W}}\left(\frac{1}{2W}\frac{\partial W}{\partial n_y} + kt\frac{\partial R^s}{\partial n_y}\right)dt \tag{E.16}$$

$$\frac{\partial^2 I_W}{\partial n_x \partial n_y} = \int_0^\infty \frac{e^{-kR^s t}}{\sqrt{W}} \left\{ \frac{1}{2W} \left[kt \left(\frac{\partial W}{\partial n_x} \frac{\partial R^s}{\partial n_y} + \frac{\partial W}{\partial n_y} \frac{\partial R^s}{\partial n_x} \right) - \frac{\partial^2 W}{\partial n_x \partial n_y} \right] \right.$$

$$\left. + kt \left(kt \frac{\partial R^s}{\partial n_x} \frac{\partial R^s}{\partial n_y} - \frac{\partial^2 R^s}{\partial n_x \partial n_y} \right) + \frac{3}{4W^2} \frac{\partial W}{\partial n_x} \frac{\partial W}{\partial n_y} \right\} \mathrm{d}t, \tag{E.17}$$

Derivatives of R_0

$$\vec{R}_0 = (x_1 - y_1, x_2 - y_2, 0),\ R_0 = \left| \vec{R}_0 \right| \tag{E.18}$$

$$\frac{\partial R_0}{\partial n_y} = -\frac{\vec{R}_0 \cdot \hat{n}_y^s}{R_0} \tag{E.19}$$

$$\frac{\partial R_0}{\partial n_x} = \frac{\vec{R}_0 \cdot \hat{n}_x}{R_0} \tag{E.20}$$

$$\frac{\partial^2 R_0}{\partial n_x \partial n_y} = -\frac{1}{R_0} \left(\hat{n}_x \cdot \hat{n}_{0y}^s - \frac{R_0 \cdot \hat{n}_x}{R_0} \frac{R_0 \cdot \hat{n}_y^s}{R_0} \right) \tag{E.21}$$

$$\hat{n}_{0y}^s = (n_{1y}, n_{2y}, 0) \tag{E.22}$$

Appendix F
Weyl–Van der Pol Approximation

Derivatives of R_g

$$R_g = \frac{\delta - \beta}{\delta + \beta} \tag{F.1}$$

The derivative in the direction of $\hat{n}_y$ is

$$\frac{\partial R_g}{\partial n_y} = \left(\frac{1}{\delta + \beta} - \frac{\delta - \beta}{(\delta + \beta)^2}\right)\frac{\partial \delta}{\partial n_y} = \frac{2\beta}{(\delta + \beta)^2}\frac{\partial \delta}{\partial n_y} \tag{F.2}$$

Similarly in the direction of $\hat{n}_x$

$$\frac{\partial R_g}{\partial n_x} = \frac{2\beta}{(\delta + \beta)^2}\frac{\partial \delta}{\partial n_x} \tag{F.3}$$

and the second derivative in the directions of $\hat{n}_x$ and $\hat{n}_y$ is

$$\frac{\partial^2 R_g}{\partial n_x \partial n_y} = \frac{2\beta}{(\delta + \beta)^2}\left(\frac{\partial^2 \delta}{\partial n_x \partial n_y} - \frac{2\beta}{\delta + \beta}\frac{\partial \delta}{\partial n_x}\frac{\partial \delta}{\partial n_y}\right) \tag{F.4}$$

Derivatives of Complex Distance ε

$$\varepsilon = \frac{1 + j}{2}\sqrt{kR^s}(\beta + \delta) \tag{F.5}$$

The derivative in the direction of $\hat{n}_y$ is

M. Ochmann and R. Piscoya, *Theory and Application of Acoustic Sources Using Complex Analysis*, https://doi.org/10.1007/978-981-33-6040-2

$$\frac{\partial \varepsilon}{\partial n_y} = \frac{1+j}{2}\left(\frac{k}{2\sqrt{kR^s}}(\beta+\delta)\frac{\partial R^s}{\partial n_y} + \sqrt{kR^s}\frac{\partial \delta}{\partial n_y}\right)$$

which can be rewritten to

$$\frac{\partial \varepsilon}{\partial n_y} = \frac{\varepsilon}{2R^s}\frac{\partial R^s}{\partial n_y} + \frac{\varepsilon}{\beta+\delta}\frac{\partial \delta}{\partial n_y} \tag{F.6}$$

Similarly in the direction of $\hat{n}_x$

$$\frac{\partial \varepsilon}{\partial n_x} = \frac{\varepsilon}{2R^s}\frac{\partial R^s}{\partial n_x} + \frac{\varepsilon}{\beta+\delta}\frac{\partial \delta}{\partial n_x} \tag{F.7}$$

and the second derivative in the directions of $\hat{n}_x$ and $\hat{n}_y$ is

$$\begin{aligned}\frac{\partial^2 \varepsilon}{\partial n_x \partial n_y} &= \frac{1}{2R^s}\left(\frac{\partial \varepsilon}{\partial n_x} - \frac{2\varepsilon}{R^s}\frac{\partial R^s}{\partial n_x}\right)\frac{\partial R^s}{\partial n_y} \\ &+ \frac{1}{\beta+\delta}\left(\frac{\partial \varepsilon}{\partial n_x} - \frac{\varepsilon}{\beta+\delta}\frac{\partial \delta}{\partial n_x}\right)\frac{\partial \delta}{\partial n_y} \\ &+ \varepsilon\left(\frac{1}{\beta+\delta}\frac{\partial^2 \delta}{\partial n_x \partial n_y} + \frac{1}{2R^s}\frac{\partial^2 R^s}{\partial n_x \partial n_y}\right)\end{aligned} \tag{F.8}$$

Derivatives of $F(\boldsymbol{\varepsilon})$

$$F = \sqrt{\pi}\varepsilon w(j\varepsilon) \tag{F.9}$$

The derivative in the direction of $\hat{n}_x$ is given by

$$\frac{\partial F}{\partial n_y} = \sqrt{\pi}\left(w(j\varepsilon) + j\varepsilon w'(j\varepsilon)\right)\frac{\partial \varepsilon}{\partial n_y}$$

The total derivative of w is known to be

$$w'(z) = -2zw(z) + \frac{2j}{\sqrt{\pi}} \tag{F.10}$$

then

$$\frac{\partial F}{\partial n_y} = \left(\left(\frac{1}{\varepsilon} + 2\varepsilon\right)F - 2\varepsilon\right)\frac{\partial \varepsilon}{\partial n_y} \tag{F.11}$$

Similarly in the direction of $\hat{n}_x$

$$\frac{\partial F}{\partial n_x} = \left(\left(\frac{1}{\varepsilon} + 2\varepsilon \right) F - 2\varepsilon \right) \frac{\partial \varepsilon}{\partial n_x} \tag{F.12}$$

The second derivative in the directions of $\hat{n}_x$ and $\hat{n}_y$ is

$$\frac{\partial^2 F}{\partial n_x \partial n_y} = \left(\frac{\partial}{\partial n_x} \left(\frac{F}{\varepsilon} + 2\varepsilon F \right) - 2 \frac{\partial \varepsilon}{\partial n_x} \right) \frac{\partial \varepsilon}{\partial n_y}$$

and after regrouping terms, we obtain

$$\begin{aligned} \frac{\partial^2 F}{\partial n_x \partial n_y} &= \left(\frac{1}{\varepsilon} + 2\varepsilon \right) \frac{\partial F}{\partial n_x} \frac{\partial \varepsilon}{\partial n_y} \\ &+ \left(\left(2 - \frac{1}{\varepsilon^2} \right) F - 2 \right) \frac{\partial \varepsilon}{\partial n_x} \frac{\partial \varepsilon}{\partial n_y} \\ &+ \left(\left(\frac{1}{\varepsilon} + 2\varepsilon \right) F - 2\varepsilon \right) \frac{\partial^2 \varepsilon}{\partial n_x \partial n_y} \end{aligned} \tag{F.13}$$

References

[Abr72] M. Abramowitz, I.A. Stegun, *Handbook of Mathematical Functions*, 9th edn. (Dover Publications, New York, 1972)

[Aki09] K. Aki, P.G. Richards, *Quantitative Seismology*, 2nd edn. (University Science Books, California, 2009)

[Att02] K. Attenborough, Sound propagation close to the ground. Annual Rev. Fluid Mech. **34**, 51–82 (2002)

[Att07] K. Attenborough, K.M. Li, K. Horoshenkov, *Predicting Outdoor Sound* (Taylor and Francis, London and New York, 2007)

[Bri09] H. Brick, Application of the boundary element method to combustion noise and half-space problems, Ph.D. thesis, Chalmers University of Technology, Göteborg, Sweden (2009)

[Bur71] A.J. Burton, G.F. Miller, The application of integral equation methods to numerical solution of some exterior boundary-value problems. Proc. R. Soc. Lond. A **323**, 201–210 (1971)

[Cha95] S.N. Chandler-Wilde, D.C. Hothersall, Efficient calculation of the Green function for acoustic propagation above a homogeneous impedance plane. J. Sound Vib. **180**(5), 705–724 (1995)

[Che10] Z.S. Chen, H. Waubke, A simple method for computing the Green functions of the helmholtz equation in the two-dimensional impedance half space. J. Comput. Acoust. **18**(1), 1–11 (2010)

[Chi75] C.F. Chien, W.W. Soroka, Sound propagation along an impedance plane. J. Sound Vib. **43**(1), 9–20 (1975)

[Cou81] M. Couture, P.-A. Belanger, From Gaussian beam, to complex-source-point spherical wave. Phys. Rev. A **24**, 355–359 (1981)

[Cul79] A.L. Cullen, P.K. Yu, Complex source-point theory of the electromagnetic open resonator. Proc. R. Soc. Lond. A **366**, 155–171 (1979)

[Dav02] B. Davies, *Integral Transforms and Their Applications*, 3rd edn. (Springer, New York, Berlin, Heidelberg, 2002)

[Des71] G.A. Deschamps, Gaussian beam as a bundle of complex rays. Electron. Lett. **7**, 684–685 (1971)

[Di93] X. Di, K.E. Gilbert, An exact Laplace transformulation for a point source above a ground surface. J. Acoust. Soc. Am. **93**, 714–720 (1993)

[Duf01] G.D. Duffy, *Green's Functions with Applications* (Chapman&Hall/CRC, Boca Raton, London, New York, Washington DC, 2001)

[Emb96] T.F.W. Embleton, Tutorial on sound propagation outdoors. J. Acoust. Soc. Am. **100**, 31–48 (1996)

[Erd54] A. Erdélyi (ed.), *Tables of Integral Transforms*, vol. I (McGraw-Hill, New York, Toronto, London, 1954)

M. Ochmann and R. Piscoya, *Theory and Application of Acoustic Sources Using Complex Analysis*, https://doi.org/10.1007/978-981-33-6040-2

[Faw00] J.A. Fawcett, Complex-image approximations to the half-space acousto-elastic Green's function. J. Acoust. Soc. Am. **108**, 2791–2795 (2000)

[Faw03a] J.A. Fawcett, A method of images for a penetrable acoustic waveguide. J. Acoust. Soc. Am. **113**, 194–204 (2003)

[Faw03b] J.A. Fawcett, Evaluation of the integrals of target/seabed scattering using the method of complex images. J. Acoust. Soc. Am. **114**, 1406–1415 (2003)

[Fel76a] L.B. Felsen, Complex-source-point-solutions of the field equations and their relation to the propagation and scattering of Gaussian beams, in *Instituto Nazionale di Alta Matematica, Symposia Matematica*, vol. 18 (Academic, London, 1976), pp. 40–56

[Fel76b] L.B. Felsen, Evanescent waves. J. Optical. Soc. Am. **66**, 751–760 (1976)

[Fel94] L.B. Felsen, N. Marcuvitz, *Radiation and Scattering of Waves, IEEE Press Series on Electromagnetic Waves* (Wiley-Interscience, New York, 1994), S. 558

[Fil99] P. Filippi, D. Habault, J.-P. Lefebvre, A. Bergassoli, *Acoustic: Basics Physics, Theory and Methods* (Academic Press, San Diego, London, Boston, 1999)

[Gor94] N. Gorenflo, An exact far-field inversion for the Born approximation and plane wave decompositions for Hankel function. ZAMP **45**, 784–793 (1994)

[Gra80] I.S. Gradshteyn, I.M. Ryzhik, *Table of Integrals, Series, and Products* (Academic Press, New York, 1980)

[Gra07] I.S. Gradshteyn, I.M. Ryzhik, *Table of Integrals, Series, and Products*, 7th edn. (Academic Press, Elsevier, San Diego, 2007).

[Guy94] J.L. Guyader, Methods to reduce computing time in structural acoustic prediction, in *Proceedings of the 3rd International Congress on Air- and Structure-Borne Sound and Vibration*, ed. by M. Crocker, vol. 1 (International Scientific Pub., Auburn, 1994), pp. 5–20

[Hab81] D. Habault, P.J.T. Filippi, Ground effect analysis: surface wave and layer potential representations. J. Sound Vib. **79**, 529–550 (1981)

[Hac84] R.H. Hackman, The transition matrix for acoustic and elastic wave scattering in prolate spheroidal coordinates. J. Acoust. Soc. Am. **75**, 35–45 (1984)

[Hei10] R.O. Hein Hoernig, Green's functions and integral equations for the Laplace and Helmholtz operators in impedance half-spaces, Thesis, Mathématique, Ecole Polytechnique X. Français.pastel-00006172 (2010)

[Hey89] E. Heyman, L.B. Felsen, Complex-source pulsed-beam fields. J. Opt. Soc. Am. A **6**, 806–817 (1989)

[Hey00] E. Heyman, V. Lomakin, G. Kaiser, Physical source realization of complex source pulsed-beams. J. Acoust. Soc. Am. **107**, 1880–1891 (2000)

[Hey01] E. Heyman, L.B. Felsen, Gaussian beam and pulsed-beam dynamics: complex-source and complex-spectrum formulations within and beyond paraxial asymptotics. J. Optical. Soc. Am. A **18**, 1588–1611 (2001)

[Hol94] F. Holste, Eine Ersatzstrahlermethode zur Berechnung des abgestrahlten Schallfeldes von Triebwerken (An equivalent source method for calculating the radiated sound field of jet engines), in *Fortschritte der Akustik* (DAGA, Dresden, 1994), pp. 773–776

[Ing51] U. Ingard, On the reflection of a spherical sound wave from an infinite plane. J. Acoust. Soc. Am. **23**(3), 329–335 (1951)

[Izm70] A.A. Izmest'ev, One parameter wave beams in free space. Radiophys. Quant. Electron. **13**, 1062–1068 (1970)

[Kai00] G. Kaiser, Complex-distance potential theory and hyperbolic equations, in *Clifford Algebras and Their Applications in Mathematical Physics. Progress in Physics*, ed. by J. Ryan, W. Sprößig, vol. 19 (Birkhäuser, Boston, MA, 2000)

[Kai01] G. Kaiser, Distributional sources for Newman's holomorphic field. https://arxiv.org/abs/gr-qc/0108041 (2001)

[Kai03] G. Kaiser, Physical wavelets and their sources: real physics in complex spacetime. J. Phys. A Math. Gen. **36**, R291–R338 (2003)

[Kis07] A.P. Kiselev, Localized light waves: paraxial and exact solutions of the wave equation (a review). Opt. Spectrosc. **102**, 603–622 (2007)

[Li96] Y.L. Li, M.J. White, Near field computation for sound propagation above ground—using complex image theory. J. Acoust. Soc. Am. **99**, 755–760 (1996)

[Lin84] I.V. Lindell, E. Alanen, Exact image theory for the Sommerfeld half-space problem. Part I: vertical magnetic dipole, 126–133. Part II: vertical electric dipole, 841–847. Part III: general formulation, 1027–1032. IEEE Trans. Antennas Propag. **32** (1984)

[Lin85] I.V. Lindell, E. Alanen, K. Mannersalo, Exact image theory for impedance computation of antennas above the ground. IEEE Trans. Antennas Propag. **33**, 937–945 (1985)

[Lin86] I.V. Lindell, E. Alanen, H.V. Bagh, Exact image theory for the calculation of fields transmitted through a planar interface of two media. IEEE Trans. Antennas Propag. **34**, 129–137 (1986)

[Lin92] I.V. Lindell, *Methods for Electromagnetic Field Analysis* (IEEE Press, New Jersey, 1992)

[Mar16] S. Marburg, The Burton and Miller method: unlocking another mystery of its coupling parameter. J. Comput. Acoust. **24**(1), 1550016(20 p) (2016)

[Mec89] F.P. Mechel, *Schallabsorber Band. I* (*Sound Absorbers, vol. I*), Chap. 13 (S. Hirzel, Stuttgart, 1989), pp. 513–666

[Mec00a] F.P. Mechel, Fundamentals of mirror and corner sources. ACUSTICA United with Acta Acustica **86**, 192 (2000)

[Mec00b] F.P. Mechel, A line source above a plane absorber. ACUSTICA United with Acta Acustica **86**, 203–215 (2000)

[Mel04] V. Mellert, B. Schwarz-Röhr, The origin of the audio signal in a beam of modulated ultrasound in air, in *Proceedings of the Joint Congress CFA/DAGA'04*, Strasbourg, France, pp. 651–652 (2004)

[Mor53] P.M. Morse, H. Feshbach, *Methods of Theoretical Physics* (McGraw-Hill, New York, 1953)

[Moh82] A. Mohsen, On the evaluation of Sommerfeld integrals. Proc. Inst. Electr. Eng. **129**, 177–182 (1982)

[Moh11a] A. Mohsen, M. Ochmann, On the boundary integral equation treatment of exterior acoustic problems. J. Comput. Acoust. **19**(4), 395–405 (2011)

[Moh11b] A. Mohsen, R. Piscoya, M. Ochmann, The application of the dual surface method to treat the non-uniqueness in solving acoustic exterior problems. Acta Acustica United with Acustica **97**(4), 699–707 (2011)

[Nor86] A.N. Norris, Complex point-source representation of real point sources and the Gaussian beam summation method. J. Optical. Soc. Am. A **3**, 2005–2010 (1986)

[Nor97] A.N. Norris, Th.B. Hansen, Exact complex source representation of time-harmonic radiation. Wave Motion **25**, 127–141 (1997)

[Och95] M. Ochmann, The source simulation technique for acoustic radiation problems. Acustica **81**, 512–527 (1995)

[Och99] M. Ochmann, The full-field equations for acoustic radiation and scattering. J. Acoust. Soc. Am. **105**, 2574–2584 (1999)

[Och00] M. Ochmann, Source simulation techniques for solving boundary element problems, in *Boundary Elements in Acoustics, Advances and Applications. Advances in Boundary Element Series*, Chap. 7, ed. by O. von Estorff (WIT Press, Computational Mechanics Publications, Southampton, Boston, 2000)

[Och04] M. Ochmann, The complex equivalent source method for sound propagation over an impedance plane. J. Acoust. Soc. Am. **116**(6), 3304–3311 (2004)

[Och05a] M. Ochmann, A boundary element formulation with complex source points for computing sound propagation over an impedance plane, in *Proceedings of the "12th International Congress on Sound and Vibration (ICSV12)"*, Lissabon, Portugal (2005)

[Och05b] M. Ochmann, Complex source point method for computing the Green's function over an arbitrary impedance plane, in *Proceedings of the NOVEM 2005*, St. Raphael, France (2005)

[Och06] M. Ochmann, R. Piscoya, The source simulation technique with complex source points for computing acoustic radiation problems, in *Proceedings of the "13th International Congress on Sound and Vibration (ICSV13)"*, Wien, Austria (2006)

[Och08a] M. Ochmann, F.P. Mechel, Analytical and numerical methods in acoustics, in *Formulas of Acoustics*, 2nd edn., ed. by F.P. Mechel (Springer, Berlin, Heidelberg, 2008), pp. 1019–1107 (see Kap. O.4)

[Och08b] M. Ochmann, R. Piscoya, The equivalent source method for computing acoustical radiation and scattering: theory and extensions, in *Proceedings of Leuven Symposium on Applied Mechanics in Engineering (LSAME 08)*, Mar 31 to Apr 2, 2008

[Och08c] M. Ochmann, H. Brick, Acoustical radiation and scattering above an impedance plane, in *Computational Acoustics of Noise Propagation in Fluids. Finite and Boundary Element Methods*, Chap. 17, ed. by S. Marburg, B. Nolte (Springer, Berlin, 2008), pp. 459–494

[Och09] M. Ochmann, R. Piscoya, Acoustic multipoles with complex source points, in *16th International Congress on Sound and Vibration (ICSV16)*, Krakow, Poland (2009)

[Och11] M. Ochmann, Closed form solutions for the acoustical impulse response over a mass like or an absorbing plane. J. Acoust. Soc. Am. **129**(6), 3502–3512 (2011)

[Och13a] M. Ochmann, Exact solutions for sound radiation from a moving monopole above an impedance plane. J. Acoust. Soc. **133**(4), 1911–1921 (2013)

[Och13b] M. Ochmann, Modelling of acoustical Green's functions above impedance planes by a superposition integral, in *Proceedings AIA-DAGA 2013*, Meran (2013)

[Och13c] M. Ochmann, R. Piscoya, Application of the acoustical boundary element method (BEM) and related Green's functions, in *20th International Congress on Sound and Vibration (ICSV20)*, July 2013, Bangkok, Thailand (2013)

[Och16] M. Ochmann, R. Piscoya, Contributions of Manfred Heckl to the equivalent source method and its extension to sources with complex singularities, in *Proceedings Manfred Heckl Memorial Session at Internoise* (2016)

[Och18] M. Ochmann, Exact solution for the acoustical impulse response of a line source above an absorbing plane. J. Acoust. Soc. Am. **144**(3), 1539–1549 (2018)

[Och20] M. Ochmann, The acoustical impulse response of a line source of finite length above an absorbing plane. J. Acoust. Soc. **147**(5), 3691–3701 (2020)

[Ols08] D. Olszewski, Stark gerichtete Audio-Beschallung mit parametrischem Ultraschall-Lautsprecher (Strongly focussed audio sound with parametric ultrasonic loudspeaker), Dissertation der Fakultät für Elektrotechnik, Informatik und Mathematik der Universität Paderborn, Diss. EIM-E/237 (available at https://d-nb.info/989972852/34, accessed Sept 2020) (2008)

[O'Ne12] M. O'Neil, L. Greengard, A. Pataki, On the efficient representation of the half-space impedance Green's function for the Helmholtz equation. arXiv:1109.6708v2 (2012)

[Pav16] G. Pavić, A. Lindberg, Airborne sound characterisation by a dummy source approach. J. Sound Vibr. **392**, 91–112 (2016)

[Pir98] R. Pirjola, A. Viljanen, Complex image method for calculating electric and magnetic fields produced by an auroral electrojet of finite length. Ann. Geopyhysicae **16**, 1434–1444 (1998)

[Pis16] R. Piscoya, M. Ochmann, Boundary elements for noise barriers above an impedance ground, in *Proceedings of the Internoise 2016*, Hamburg, Germany (2016)

[Pis17] R. Piscoya, M. Ochmann, Acoustical Green's function and boundary element techniques for 3D half-space problems. J. Comput. Acoustics **25**(04), 1730001 (2017)

[Pis19] R. Piscoya, M. Ochmann, Implementation of different formulas for special Green's functions in a 3D half-space BEM. J. Comput. Acoustics **1**, 1850051 (2019)

[Ras89] R. Raspet, G.E. Baird, The acoustic surface wave above a complex impedance ground. J. Acoust. Soc. Am. **85**, 638–640 (1989)

[Rud47] I. Rudnick, The propagation of an acoustic wave along a boundary. J. Acoust. Soc. Am. **19**(2), 348–356 (1947)

[Sch68] H.A. Schenk, Improved integral equation formulation for acoustic radiation problems. J. Acoust. Soc. Am. **44**, 41–58 (1968)

[Sey88] A.F. Seybert, B. Soenarko, Radiation and scattering of acoustic waves from bodies of arbitrary shape in a three-dimensional half space. Trans. ASME **110**, 112–117 (1988)

[Shi77] S.Y. Shin, L.B. Felsen, Gaussian beam modes by multipoles with complex source points. J. Opt. Soc. Am. **67**, 699–700 (1977)

[Shu93] R.M. Shubair, Y.L. Chow, A simple and accurate complex image interpretation of vertical antennas present in contiguous dielectric half-spaces. IEEE Trans. Antennas Propag. **41**, 806–812 (1993)

[Smy68] W.R. Smythe, *Static and Dynamic Electricity*, 3rd edn. (McGraw-Hill, New York, 1968). Cited in https://en.wikipedia.org/wiki/Oblate_spheroidal_coordinates

[Sku71] E. Skudrzyk, *The Foundations of Acoustics, Basis Mathematics and Basis Acoustics* (Springer, Wien, New York, 1971)

[Som09] A. Sommerfeld, Über die Ausbreitung der Wellen in der drahtlosen Telegraphie ("The propagation of waves in wireless telegraphy"). Ann. Phys. **28**, 665–736 (1909)

[Som26] A. Sommerfeld, Über die Ausbreitung der Wellen in der drahtlosen Telegraphie ("The propagation of waves in wireless telegraphy"). Ann. Phys. **81**, 1135–1153 (1926)

[Som78] A. Sommerfeld, *Partielle Differentialgleichungen Der Physik (Partial Differential Equations in Physics)*, vol. VI (Harri Deutsch, Thun-Frankfurt/M, 1978)

[Tag09] A.M. Tagirdzhanov, A.S. Blagovestchenskii, A.P. Kiselev, "Complex source": singularities in real space, in *Progress in Electromagnetics Research Symposium Proceedings*, Moscow, Russia, 18–21 Aug 2009, pp. 1527–1529

[Tag11] A.M. Tagirdzhanov, A.S. Blagovestchenskii, A.P. Kiselev, 'Complex source' wavefields: sources in real space. J. Phys. A Math. Theor. **44**, 425203(15 p) (2011)

[Tag13] A.M. Tagirdzhanov, A "complex source" in the 2D real space. J. Math. Sci. **194**, 98–103 (2013)

[Tag15] A.M. Tagirdzhanov, A.S. Blagovestchenskii, A.P. Kiselev, Time-harmonic wavefields of "complex sources" and their sources in the real space. J. Math. Sci. **206**, 315–327 (2015)

[Tar05a] G. Taraldsen, A note on reflection of spherical waves. J. Acoust. Soc. Am. **117**, 3389–3392 (2005)

[Tar05b] G. Taraldsen, The complex image method. Wave Motion **43**, 91–97 (2005)

[Tho76] S.I. Thomasson, Reflections of waves from a point source by an impedance boundary. J. Acoust. Soc. Am. **59**(4), 956–963 (1974)

[Tyg87] M. Tygel, R. Hubral, Transient waves in layered media, in *Methods in Geochemistry and Geophysics,* vol. 26 (Elsevier, Amsterdam, Oxford, New York, Tokyo, 1987)

[Vla00] N. Vlahopoulos, Indirect variational boundary element method in acoustics, in *Boundary Element Acoustics Fundamental and Computer Codes*, Chap. 6 (WIT Press, London, 2000)

[Vit96] St. Vitebskiy, K. Strugess, Short-pulse plane-wave scattering from buried perfectly conducting bodies of revolution. IEEE Trans. Antennas Propag. **44**, 143–151 (1996)

[Wen74] A.R. Wenzel, Propagation of waves along an impedance boundary. J. Acoust. Soc. **55**(5), 956–963 (1974)

[Wes63] P.J. Westervelt, Parametric acoustic array. J. Acoust. Soc. Am. **35**, 535–537 (1963)

[Wey19] H. Weyl, Ausbreitung elektromagnetischer Wellen über einem ebenen Leiter ("Propagation of electromagnetic waves above a flat conductor"). Ann. Phys. **60**, 481–500 (1919)

[Yan91] J.J. Yang, Y.L. Chow, D.G. Fang, Discrete complex images of a three-dimensional dipole above and within a lossy ground, in *IEE-Proceedings-H*, vol. 138 (1991), pp. 319–326

[Yon83] M. Yoneyama, J. Fujimoto, The audio spotlight: an application of nonlinear interaction of sound waves to a new type of loudspeaker design. J. Acoust. Soc. Am. **73**, 1532–1536 (1983)

[Zau86] E. Zauderer, Complex argument Hermite-Gaussian and Laguerre-Gaussian beams. J. Opt. Soc. Am. A **3**, 465–469 (1986)

[Zem65] A.H. Zemanian, *Distribution Theory and Transform Analysis* (McGraw-Hill, New York, 1965)

[Zio85] R.W. Ziolkowski, Exact solutions of the wave equation with complex source locations. J. Math. Phys. **26**, 861–863 (1985)

Internet Links

[WikOs] https://en.wikipedia.org/wiki/Oblate_spheroidal_coordinates. Accessed Jan 2021
[WikWD] https://en.wikipedia.org/wiki/Weak_derivative. Accessed Jan 2021
[WikHa] https://en.wikipedia.org/wiki/Hankel_transform. Accessed Jan 2021

Index

M. Ochmann and R. Piscoya, *Theory and Application of Acoustic Sources Using Complex Analysis*, https://doi.org/10.1007/978-981-33-6040-2